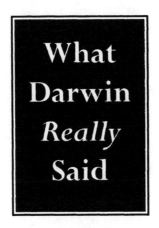

What
Darwin
Really
Said

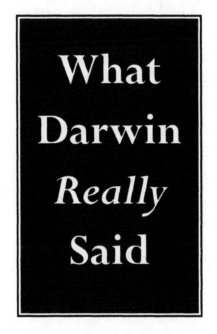

What
Darwin
Really
Said

BENJAMIN
FARRINGTON

Foreword by
Stephen Jay Gould

SCHOCKEN BOOKS
NEW YORK

Library of Congress Cataloging in Publication Data
Farrington, Benjamin, 1891–1974
What Darwin really said.
(What they really said series)
Bibliography.
Includes index.
1. Darwin, Charles, 1809–1882. 2. Naturalists—England—
Biography. I. Title.
QH31.D2F34   1982   575'.0092'4[B] 82-5557   AACR2
ISBN: 978-0-8052-1062-0

Manufactured in the United States of America

145052501

Acknowledgments

I have to thank Mr. Leonard Barnes for permission to quote from his poem, *The Homecoming* (London: Peter Davies, 1961), and Messrs. William Collins, Sons & Co. Ltd., and the editor for extracts from *The Autobiography of Charles Darwin,* 1809–1882, edited by Nora Barlow.

Contents

Contents

In Praise of Charles Darwin

Foreword by Stephen Jay Gould

As if to provide future revelers with a double excuse for rejoicing, Charles Darwin was kind enough to publish his great work, *On the Origin of Species,* when he was exactly fifty years old. In 1959 we could celebrate both the centennial of his book and the sesquicentennial of his birth. And celebrate we did, with symposia and conferences throughout the world.

The only sour notes amidst all this merriment were sounded by scholars who followed a tradition, then prevalent, of labeling Darwin an anomaly in the history of ideas. He was, they said, a slow man of ordinary skills who happened to be in the right place at the right time; zeal and patience were the only virtues they acknowledged. One well-known biographer described Darwin as "limited intellectually and insensitive culturally"; another judged him "a great assembler of facts and a poor joiner of ideas . . . a man who does not belong with the great thinkers."

The 1959 celebrations inspired a wave of Darwinian scholarship that put these deprecating interpretations firmly to rest and elucidated the different but genuine nature of Darwin's genius. Thus, as the next great excuse for a celebration arises—Darwin died one hundred years ago this April, and we must mark deaths as well as births in our uncertain world—we see a different Darwin, a much more powerful and admirable man.

Darwin has been the inspiration of my life and work, joining my father and Joe DiMaggio in the select trio of men who most profoundly influenced my life. For the centennial of his death, I write this appreciation—a frankly celebratory essay, but an honest one. Let us rejoice that we can identify, in our complex and ambiguous world, a man with such power of thought and such influence upon us all—a man who, at the same time, managed to be an exemplary human being.

Darwin was born in Shrewsbury in 1809, on the very day that witnessed the birth of Abraham Lincoln. He had all the advantages that social class and money could offer. His father was a wealthy physician, his grandfather Erasmus Darwin, a celebrated writer whose books about nature, in heroic couplets, are often (and mistakenly) read as harbingers of his grandson's views. He was an indifferent student at Cambridge, primarily because of a lack of commitment. But Darwin suffered a sea change aboard the *Beagle* and returned, after five years of traveling around the globe, a confirmed naturalist, having abandoned his earlier plans to become a country parson. According to popular legend, the finches and tortoises of the Galapagos pushed Darwin over the edge to heresy. In fact, they (and other observations he made while aboard the *Beagle*) only nudged him in that direction; two years of concentrated effort in London after the voyage fueled his transition. He arranged his notes, read voraciously in all fields of science, poetry, and philosophy, filled notebook after notebook with telegraphic insights, and finally, in 1838, put it all together in the theory of natural selection.

About the externals of the rest of his life, we can say rather little of conventional interest. He married his cousin Emma Wedgwood and lived a long and happy life with her, untouched by the slightest breath of poverty or scandal. He never again left the British Isles, and rarely even ventured forth from his country house at Downe, on the outskirts of London. His tragedies were those of all his contemporaries—the early deaths of several beloved children. His daily trials were retching and flatulence— a chronic illness of unknown cause that has provided endless (and largely fruitless) debate among Darwinian scholars.

But ah, consider the turmoil within his mind! For thirty years he sat at Downe, turning out book after book, some disarmingly obscure (including tomes on climbing plants, orchids, and the formation of vegetable mold by earthworms), but all part and parcel of a general theory and approach that revolutionized human thought. Few men have influenced the world so profoundly, and from such a citadel of apparent calm.

Why did Darwin, rather than Jean Baptiste Lamarck or Robert Chambers or any of the numerous evolutionists who preceded or followed him during the nineteenth century, become such a symbol and prime mover of the greatest transition in the history of biological thought? Scientists and historians have pondered this riddle for a century, and have not resolved it thoroughly (thereby leaving enough work for the bicentennial in 2009). The points that seem important to me can be arranged in five categories, providing a framework for this appreciation.

The Usable Character of Darwin's Theory

Darwin's fame cannot lie merely in the fact of his evolutionary convictions, for he had several predecessors among the greatest scientists of Europe (Lamarck and Geoffroy de Saint-Hilaire in France, in particular). But these predecessors had developed speculative theories, largely devoid of direct evidence and not subject to fruitful testing. Science can only traffic in usable and operational ideas.

On the Origin of Species provides copious evidence and direct suggestions for research, not merely a cosmic, untestable view suited more for contemplation in awe (or disgust) than for immediate use, scrutiny, and extension. Darwin, for the first time, gave scientists something practical to do.

Lamarck's theory had proposed an inherent force that "tends incessantly to complicate organization." He contrasted this internal perfecting tendency with the "influence of circumstances," or what we could today call adaptation to local environment. For Lamarck, small changes that could be observed and manipulated were not the substance of evolution's most important process—the drive toward perfection—but only tangential deflections that adapted creatures to local circumstances.

In Darwin's theory of natural selection, on the other hand, these small changes *are*, by extension, all of evolution. Darwin dispenses with unknown internal forces, and attempts to render all change on all scales as the accumulated product of small and observable modifications. Thus the breeding of pigeons, and minor geographic variation within natural species, become the stuff of all evolution. In studying what can be observed and measured, scientists examine the essence of the process, not merely a deflecting force opposed to an unknown internal drive. Darwin made evolution a *workable* science. From a professional's standpoint, there can be no greater praise (I myself believe that Darwin went too far in attempting to reduce all large-scale phenomena to the gradual accumulation of small changes under natural selection. But this is a subject for another time, and my major point is only further confirmed: if Darwin had not established a *workable* theory of evolution, we would still be spinning stories in armchairs or over cocktails, not developing testable critiques.)

The Radical Implications of Natural Selection

The common denominator of evolutionary theories proposed by Darwin's rivals lies in their congeniality with many traditional biases of Western thought that Darwin was trying to challenge or strip away. They view evolution as a foreordained process ruled by principles of inherent progress.

Natural selection, however, is a theory of local adaptation only. Changes that, in our anthropocentric way, we choose to call progressive represent only one pathway of adaptation to changing local environments. Every large-brained mammal harbors species of parasites so morphologically "degenerate" that they are little more than bags of reproductive tissue. Yet who can say that one or the other is "better" or any surer of evolutionary persistence?

If a denial of inherent progress were not radical enough, Darwin also introduced the specter of randomness into evolutionary theory. To be sure, randomness only provides a source of *variation* in Darwin's theory. Natural selection (a deterministic process) then scrutinizes the spectrum of random variants and preserves those individuals best adapted to changing local environments. Still, chance in any form was anathema to many nineteenth-century thinkers, both then and now.

Darwin's theory also challenged the comforting assumption that evolution must be purposive, working toward the good of species or ecosystems. The theory of natural selection, established in perhaps unconscious analogy to the individualistic, laissez-faire economics of Adam Smith (whom Darwin had been studying intensely just before he formulated his theory), speaks only of individuals struggling for personal success. In modern terms, natural selection concerns the unconscious struggle of individuals to leave more of their genes in surviving offspring. Any benefits to species, any harmony in ecosystems, arise merely as a by-product of this struggle among individuals, or, in the case of ecosystems, as a natural balance among competitors.

What then of spirit, of vital forces, of God himself? No intervening spirit watches lovingly over the affairs of nature (though Newton's clock-winding god might have set up the machinery at the beginning of time and then let it run). No vital forces propel evolutionary change. And whatever we may think of God, his existence is not manifest in the products of nature.

Darwin was not an atheist. He probably retained a belief in some kind of personal god—but he did not grant his deity a directly and continuously intervening role in the evolutionary process. Many have viewed this message as pessimistic, or even nihilistic. I have always understood it (as I believe Darwin intended) as positive and exhilarating. It teaches us that the meaning of our lives cannot be read passively from the works of nature, but that we must struggle, think, and construct that meaning for ourselves. Moreover, Darwin maintained deep humility before the difficulty of such a task. He understood the limits of science.

The Universal Scope of Darwin's Vision

Many of Darwin's fellow evolutionists, either through lack of courage or adherence to tradition, constructed tortured arguments to exclude human beings from their system and to make a divine exception for this one peculiar primate. Darwin persevered, and built a general theory applicable to all organisms. We can, I believe, discern a trilogy of increasing daring among his major works. He published the *Origin of Species* and established his general theory in 1859. Of our species, he said only: "Light will be thrown on the origin of man and his history"; subsequent editions ventured the slightly intensified "much light." In 1871 he published *The Descent of Man*, and argued that our bodies had also been molded by the forces of natural selection. Finally, in *The Expression of the Emotions in Man and Animals* (1872), he dared claim that our most refined and most particularly human behavior—the expression of our emotions—also reflected an evolutionary past. We express disgust with a facial motion associated with the adaptive act of vomiting. We curl our lips in rage, raising them most just where our useless canine teeth protrude; yet in our ancestors, these very teeth were long and sharp weapons. In our soul as well as our body, we display "the humble stamp of a lowly origin."

The Consistency and Depth of Darwin's Thought

Charles Darwin wrote 15 books (excluding four monographs on the taxonomy of barnacles and his contribution to Captain Fitzroy's narrative of the *Beagle*'s voyage). Traditionally, these books were regarded as a motley collection—a few revolutionary tomes to be sure, but mostly the trivial playthings of a doddering naturalist. How else to view a book on "the structure and distribution of coral reefs," or "on the various contrivances by means of which orchids are fertilized by insects," or on "the formation of vegetable mould, through the action of worms"?

I think Darwinian scholars would now agree, however, that the entire corpus of his work is one consistent, ramifying, and remarkable exploration of his new view of life and its consequences. All his books are either about evolution or about the extension to other subjects of his evolutionary method (the study of historical continuity and the criteria for inferring genealogical connection). His theory of coral reefs, for example, depends upon a recognition that all the varied forms of modern reefs can be understood as stages of a single historical sequence (continued upgrowth of reefs as supporting islands sink below the sea). This theory, strongly challenged in Darwin's time, is now abundantly affirmed; the method of argument is identical with the inference of evolution from different stages in the process of speciation displayed by modern populations.

The orchid book is not a compendium of a hobbyist's minutiae, but a long argument about why the imperfection of organic design illustrates evolutionary descent. When environments change, organisms must modify ancestral parts for new functions. This legacy of the past precludes the development of optimum designs. Orchids entice insects by modifying the ordinary parts of flowers for new roles. The worm book treats Darwin's favorite evolutionary theme; that an accumulation of small changes produces large effects in the long run.

Darwin also recognized the profound and tumultuous effect that evolution would work upon other traditional disciplines far from science. In a remarkable passage from an early notebook, for example, Darwin cuts through two thousand years of philosophical tradition with a single phrase: "Plato says in *Phaedo* that our imaginary ideas arise from the pre-existence of the soul, are not derivable from experience—read monkey for pre-existence."

True Heroes Must Be Made of Flesh and Blood

Had Darwin been a cold fish, or a nasty, exploitative man, we might be less attracted to him, though we would still admire the power of his thought. Yet he was a person whose basic kindness and decency defy the numerous attempts of detractors to demean or defame him. The external calm of his life belied an inner turmoil, but Darwin suppressed his anxieties, or channeled them into illness or work (depending upon your preference in psychoanalytic theories), and remained a truly eminent Victorian.

Darwin wins us, first of all, by his fine writing. He was not, as were the scientists Thomas Henry Huxley and Charles Lyell, a powerful and elegant stylist. He often wrote pages of adequate, but ordinary, descriptive prose. But he had a flair for metaphor and occasional outbursts of controlled passion. And these gems shine all the more because they are embedded in ordinary prose and we meet them unexpectedly, with delight. Consider his metaphors of the wedge, the tangled bank, the tree of life, or my favorite (in drawing a contrast between the superficial harmony of ecosystems and the underlying struggle for existence among individuals): "We behold the face of nature bright with gladness. . . ."

When Darwin uses this power of prose to advance social positions that most of us view as enlightened (Darwin was, in nineteenth-century parlance, a liberal committed to the abolishment of restraints upon the expression of human potential), the effect can be stunning. Consider this passage on slavery from the conclusion of his *Voyage of the Beagle:*

Near Rio de Janeiro I lived opposite to an old lady, who kept screws to crush the fingers of her female slaves. I have stayed in a house where a young household mulatto, daily and hourly, was reviled, beaten, and persecuted enough to break the spirit of the lowest animal. I have seen a little boy, six or seven years old, struck thrice with a horse-whip (before I could interfere) on his naked head, for having handed me a glass of water not quite clean . . . And these deeds are done and palliated by men, who profess to love their neighbors as themselves, who believe in God, and pray that his Will be done on earth! It makes one's blood boil, yet heart tremble, to think that we Englishmen and our American descendants, with their boastful cry of liberty, have been and are so guilty.

Yet I do not wish to portray Darwin as a man of cardboard, for then he would be dull in his one-dimensionality, however admirable. Darwin had faults aplenty, even if most were the common attitudes of his age. Don't think that his many favorable words about blacks reflect a general egalitarian perspective. Virtually no white male of that era—neither Franklin, nor Jefferson, nor Lincoln—doubted the innate superiority of his race. Darwin spoke well of blacks because he had respect for this race as well, but read his words about the people of Tierra del Fuego if you wish to encounter the conventional intolerance of his age. As for women, his kindness includes little evidence of respect for their intellectual potential. In one particularly chilling, though uncharacteristic, private jotting, he had this to say about the advantages of marrying: "Constant companion, (friend in old age) who will feel interested in one, object to be beloved and played with—better than a dog anyhow—Home and someone to take care of house—Charms of music and female chitchat. These things good for one's health."

Darwin's humanity, with all its foibles, shines through in his life and writing. We can feel the pain of his inner conflict when, after working for twenty years on the theory of natural selection, he receives a short note from the English naturalist Alfred Russel Wallace containing the identical theory devised one night during a malarial fit. Can he publish and preserve his legitimate priority, or must he stand aside? Gutsy humanity, or abstract virtue? He writes to Lyell, exhorting him between the lines to find an honorable yet advantageous way: "I should be extremely glad now to publish a sketch of my general views . . . but I cannot persuade myself that I can do so honorably . . . I would far rather burn my whole book, than that [Wallace] or any other man should think that I have behaved in a paltry spirit . . . My good dear friend, forgive me. This is a trumpery letter, influenced by trumpery feelings." (Lyell and other friends took the hint and arranged to publish Wallace's note along with parts of a sketch that Darwin had written during the 1840s. Author Arnold Brackman has recently suggested that Darwin actually cribbed parts of his theory from Wallace, but details of dates and places invalidate the claim.)

In any case, this debate only concerns a subsidiary issue called the principle of divergence. Priority for natural selection—the essence of the theory—cannot be denied to Darwin. He developed the idea in 1838 when Wallace was a teenager. In this, as in so many other incidents of Darwin's life, we understand in the most direct and poignant manner that science is quintessentially, a human endeavor. Let us praise famous men for this fundamental lesson.

Darwin died in April 1882. He wished to be buried in his beloved village, but the sentiment of educated men demanded a place in Westminster Abbey beside Isaac Newton. As his coffin entered the vast building, the choir sang an anthem composed for the occasion. Its text, from the

Book of Proverbs, may stand as the most fitting testimony to Darwin's greatness: "Happy is the man that findeth wisdom, and getteth understanding. She is more precious than rubies, and all the things thou canst desire are not to be compared unto her."

February 1982

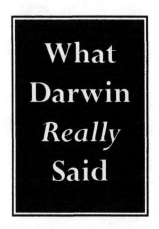

**What
Darwin
Really
Said**

Introduction

Charles Darwin, the author, among many other books, of *The Origin of Species* (1859) and of *The Descent of Man* (1871), was the central figure in a great revolution of thought which was taking place about a hundred years ago. At that time it was commonly believed that the world was only about 6,000 years old; that it had been created by God about 4,000 B.C. as a going concern; and that now, about 6,000 years later, it was still much the same as He had left it when He rested from His six days' work.

A contrary opinion had been held by some thinkers in antiquity and was now being revived with much fresh argument. This was, that the world was much more than 6,000 years old; and that the world and all the creatures on it had undergone many great changes and would undergo more. The world, in short, was not simply a going concern, but a growing concern. Moreover, it was claimed, science had begun to arrange in order the various stages of development and to determine the laws of its growth.

This theory of evolution rather than a once-for-all act of creation had been in the air for some time before Charles Darwin wrote. His own grandfather, Erasmus Darwin, was one of its most ardent and persuasive advocates. But the majority of educated men, not only churchmen but scientists, were against the idea until Charles Darwin formulated the theory in a new way and supported it with a mass of evidence which won it general acceptance.

The extent of the opposition by churchmen created the image of a clash between religion and science which has left its mark on our history. It is not forgotten how Bishop Wilberforce at a meeting of the British Association in 1860 asked indignantly whether a man who accepted an ape for his grandfather must also accept an ape for grandmother. Thomas Huxley caustically replied that for his part he would rather be descended from a humble monkey than from a man who employed his eloquence misrepresenting earnest men who were wearing out their lives in the search for truth. But there were open-minded men in the Church as well as outside it. In the same year the New Testament scholar Hort wrote to his friend Westcott: "Have you read Darwin? . . . I am inclined to think it unanswerable. In any case it's a treat to read such a book." Other churchmen too were soon "thanking God that the scientific men have shattered the idol of an infallible book."

Part of our business, then, will be to set forth the evidence advanced by Darwin which shattered the creationist views taught by the Church on the basis of the first two chapters of *Genesis*. Also we shall have to glance at what Darwin did not know. For so rapid has been the advance of knowledge that for the modern evolutionist much of what Darwin wrote is irredeemably out-of-date. Also, like everybody else, he had his blind spots.

Youth

Charles Darwin (1809–1882) was a born naturalist if ever there was one. In books he took but a languid interest. But put him face to face with nature and he came instantly alive. A sentence from the last paragraph of his *Origin of Species* is an unconscious self-portrait. "It is interesting," he writes, in his simple unpretentious style, "to contemplate a tangled bank, clothed with many plants of many kinds, with birds singing on the bushes, with various insects flitting about, and with worms crawling through the damp earth, and to reflect that these elaborately constructed forms, so different from each other, and dependent upon each other in so complex a manner, have all been produced by laws acting around us." There we have a picture of the man and his work—his fascination with nature, and not simply with the appearances of nature but with the laws by which it works.

When he was about ten years old he was sent to Shrewsbury School. There is something a little dim and depressing about his childhood. His mother had died the year before. It might be thought that he was old enough to retain many memories of her. In fact he retained few. His father was now his sole parent. Him he seems to have respected rather than loved. At school, as was the fashion of the day, he was taught nothing but Latin and Greek and a little ancient history. Being both intelligent and docile he did as he was bid and was soon learning by heart forty or fifty lines of Homer or Virgil every day. But he forgot them at once. So things went on for some six or seven years. His only refuge was his passion for collecting specimens—animal, vegetable, or mineral; that, and the chemical laboratory which his elder brother had improvised in the tool-shed at the bottom of their garden. In this way he educated himself, and what he learned in this way stuck.

His father also helped in the formation of the budding genius, though without any clear foresight of the direction his son was destined to take. Robert Darwin, son of the famous Erasmus, was himself remarkable enough. A busy family physician in Shrewsbury, he had distinguished himself sufficiently in science to have been elected a Fellow of the Royal Society. He thought he saw a doctor in Charles and, when occasion offered, took him with him on his rounds. Such an apprenticeship to the art of medicine was still possible in those less formal days. It is even said that Charles, while still in his 'teens, practised a bit on his own and had as many as a dozen patients on his list! In 1825, when he was sixteen, his father took him from school and sent him off to Edinburgh to study medicine.

Here the lectures were dull. But Charles had a real flair for medicine and might have succeeded but for two things. First, his father was now doing so well that Charles began to think he might be left independent and so could devote himself wholly to collecting! He had already read two original papers to a natural history society in Edinburgh and he was paying a Negro then living in Edinburgh to teach him to stuff birds. "I used often to sit with him," Darwin records, "for he was a very pleasant and intelligent man." The second thing was that anaesthetics had not yet been invented. The sight of two operations performed on children without chloroform was more than Charles could bear. It haunted him for years. His medical career was over and his father recalled him from Edinburgh and put before him plans for a new career.

The new career was a surprising choice and seems to indicate that Robert had no high respect for his son's intelligence. For two generations now the Darwin men had been unbelievers. They were philosophical deists who thought Christianity was for women and children. Robert now proposed that Charles should enter the Church and Charles raised no objections. He obediently proceeded to brush up his forgotten Greek, and in 1828 set off for Cambridge and the three happiest years of his life. "Upon the whole," he says in the *Autobiography* he wrote in his old age, "the three years I spent at Cambridge were the most joyous in my happy life; for I was in excellent health, and almost always in high spirits." But together with the high spirits, the outdoor sports, the conviviality and the companionship of these years, the intellectual influences that were brought to bear upon him at this time are so important for the understanding of his theory of evolution that we must devote the next chapter to sorting them out.

Cambridge:
Conflicting Influences

The mental background of Charles Darwin is not without its puzzling side. We have said that his grandfather Erasmus Darwin (1731–1802) was a pioneer of evolutionary theory. In his book *Zoonomia, or the Laws of Organic Life,* published in 1794, he had written:

Would it be too bold to imagine that in the great length of time since the earth began to exist, perhaps millions of ages before the commencement of the history of mankind, would it be too bold to imagine that all the warm-blooded animals have arisen from one living filament, which THE GREAT CAUSE endued with animality, with the power of acquiring new parts, attended with new propensities, directed by irritations, sensations, volitions and associations; and thus possessing the faculty of continuing to improve by its own inherent activity, and of delivering down those improvements by generation to its posterity, world without end!

Here we have not only a brilliant early formulation of the theory of biological evolution, plus the deistic Great Cause instead of God; we have also the suggestion that the earth had existed for "millions of ages" before the appearance of man.

This speculation as to the great antiquity of the earth was fed by the young science of geology. James Hutton (1726–1797), the Edinburgh scientist, a trained chemist and mineralogist, had become possessed with the urge not merely to describe and classify minerals but to trace their origin. His idea was that all rocks now visible on the surface of the earth had been formed out of the waste of older rocks. Exposed rocks are continually worn down by the gradual action of natural agencies—heat and cold, wind and rain, rivers and ice. The debris is carried to the sea and consolidated again under pressure, thus forming the strata of the aqueous rocks. These again are lifted to the surface by upheavals of the ocean bed and sometimes disrupted by the upthrust of molten igneous rocks from the deeper levels of the earth. When they reach the surface the cycle of disintegration, sedimentation, and so on begins again.

The connection of geology with the theory of biological evolution is strongly evident in the work of another pioneer, William Smith (1769–1839). He was a civil engineer much busied in the new enterprise of canal construction. The canal cuts revealed to him again and again the same strata in the same order lying on top of one another in widely distant places "like slices of bread and butter on a breakfast plate," to use his own homely illustration. He observed, too, that the strata could be identified by the fossils, that is by the embedded organic remains, revealing the types of life that existed when the strata were laid down. This contribution to science has won him the name of "Stratum Smith."

But while the practical men were creating the new science which was to multiply a million-fold the estimated age of the earth, the *Encyclopedia Britannica* was still discussing the problem in terms of what the Book of Genesis was supposed to mean. Had the earth been created in 4,305 B.C. or in 4,000 B.C.? Was it now 6,096 years old or not yet quite 6,000? "Be that as it may," the article concludes, "the whole account of the creation rests on the truth of the Mosaic history, which we must of necessity accept, because we can find no other which does not either abound with the grossest absurdities, or lead us into absolute darkness." The delightful and learned poet William Cowper, author of John Gilpin, was of the same opinion as the *Encyclopedia*. He was contemptuous of the new geology.

> Some drill and bore
> The solid earth, and from the strata there
> Extract a register, by which we learn
> That He who made it, and revealed its date
> To Moses, was mistaken in its age.
>
> *The Task,* Bk. II, 150–4.

In this dispute where did Darwin stand? He had read his grandfather's book and he had acquired in Edinburgh already a passion for geology. Yet in his *Autobiography* he tells us that, on setting out for Cambridge to prepare himself for the ministry, he did not then "in the least doubt the strict and literal truth of every word in the Bible." Obviously in his three carefree undergraduate years at Cambridge it did not worry Darwin to hold two contradictory views at the same time.

This explains his enthusiasm for each of the chief exponents of the two rival views, the theologian Paley and the geologist Lyell. William Paley (1743–1805), the able and persuasive author of the famous *Natural Theology or Evidences of the Existence and Attributes of the Deity collected from the Appearances of Nature* (1802), was the defender of the orthodox view of a once-for-all act of creation. Darwin read him with delight and found his logic as cogent as that of Euclid. Lyell, the first volume of whose *Principles of Geology* came out in 1830, took the opposite view. He described his book as "An Attempt to explain the Former Changes of the Earth's Surface by Reference to Causes now in Operation." The present state of the earth, according to Lyell, is the result of a slow process of change. He accepts the classification of the rocks, according to the fossilized plant and animal remains found in them, as: (1) Palaeozoic, i.e., containing older forms of life which are all marine or reptilian; (2) Mesozoic, i.e., those containing middle forms of life, when birds and mammals appear; (3) Cenozoic, i.e., those containing new forms of life in which mammals predominate. Then he himself proceeds to subdivide the Cenozoic period into three: (i) Eocene, or the dawn of the new forms of life; (ii) Miocene, when the new forms of life are numerous but still a minority of the whole; and (iii) Pliocene, when the new forms of life outnumber the old. It was Lyell's principles that were destined to prevail with Darwin; and, while orthodox opinion had declared in the words of John

Wesley (in 1770), "Death is never permitted to destroy the most inconsiderable species," Darwin in his first published book, *The Voyage of the* Beagle, was stoutly to declare: "Certainly no fact in the long history of the world is so startling as the widespread and repeated extermination of its inhabitants."

What we shall next have to consider is the nature of the evidence which destroyed, first in Darwin's mind, then in the minds of educated people generally, the validity of Paley's arguments.

The Voyage
of the "Beagle"

"During my last year in Cambridge," wrote Darwin in his *Autobiography,* "I read with care and profound interest Humboldt's *Personal Narrative.*" In this book the great German traveller tells how, in his five years of travel in South America (1799–1804), he had been able to contribute to the advance of the sciences of geology, physical geography, and mineralogy. The book, writes Darwin, "stirred up in me a burning zeal to add even the most humble contribution to the noble structure of Natural Science." Now suddenly, out of the blue, the chance was offered to him to follow in Humboldt's train. The British Admiralty had in 1820 launched a frigate, H.M.S. *Beagle,* designed from the first for scientific research. In 1831 the *Beagle* was being prepared for a five-year expedition "the object of which," Darwin records, "was to complete the survey of Patagonia and

Tierra del Fuego" begun on an earlier voyage, and "to survey the shores of Chile, Peru, and of some islands in the Pacific; and to carry a chain of chronometrical measurements round the world." Darwin's zeal and competence must have impressed his scientific acquaintances far more deeply than he had any inkling of, for the divinity student unexpectedly found himself invited to take charge of the natural history side of the expedition. Darwin jumped at the chance; but not so his father. Robert had already seen his son fail to qualify for one profession and was not now prepared to have him risk failure in another. Charles, always the dutiful son, was ready to bow to parental authority. But a wise uncle intervened, and the unenthusiastic divinity student got his chance to make his "humble contribution to the noble structure of natural science."

So it came about that in December 1831 a desperately seasick Darwin lay in his hammock in his small cabin tossing his way down the Channel. Physically he was not very happy. He was, in fact, so ill that he often wondered whether the protracted invalidism of his later years did not owe its origin to this grim experience. He shared a small cabin fifteen feet by ten with two officers. Space was found to sling his hammock—he was a big tall man—only by removing one of the drawers from the chest which stood against the wall. He had to be very tidy to find room for his personal belongings and his scientific gear in so confined a space. But after ten days the worst of his troubles were over. The *Beagle* came to port at St. Iago in the Cape Verde Islands. He had read the description of them in Humboldt, but the reality surpassed his expectations. "It has been for me a glorious day," he recorded in the diary of his voyage, "like giving to a blind man eyes. Such are my feelings, and such may they remain."

Such they did remain; and, if we are to understand the zeal and excitement which bore him up through the hardships and exertions of the next five years, we must appreciate the revolutionary novelty of Lyell's endeavour "to explain the former changes of the earth's surface by reference to causes now in operation." Men had been accustomed to explain everything in terms either of the biblical creation or of the biblical flood. Everything was still as it had always been except for a few catastrophes. Thus, to take one example, rivers were regarded as part of the permanent structure of the earth, fed with waters from underground channels. It was not known that the water in rivers is rain from the catchment area; nor was it known that rivers had carved out, and were still carving out, their own valleys. In his Edinburgh and Cambridge days Darwin had served his apprenticeship to this new science. Great Britain has the advantage for the geologist of packing a great variety into a narrow compass. To his collections of plant and animal life he had added his collection of minerals and fossils. There are stories from these days that illustrate his zeal. Tearing the bark off a decaying tree he found three different beetles at once. When he had one beetle in each hand, and the third seemed likely to escape, he tried the expedient of putting one for safe keeping into his mouth; but it excreted such a pungent liquid that he came near to losing them all. At another time in Shrewsbury an old man drew his attention to an erratic boulder. There was not another rock of the same kind nearer than Cumberland, said the old man, adding that the world would end before anybody explained how the stone got there. Darwin was not the less filled with excite-

ment because he thought he knew the answer. The action
of glaciers in transporting boulders was beginning to be
understood. But now he was to have the opportunity of
extending his observations over a vast area of land and
sea. The world was his oyster. He was soon to prove how
well equipped he was to open it.

The Voyage of the Beagle, the first and the most readable
of his many books, is one long record of his observations in
various fields, fascinating for its infectious intellectual enthusiasm and for its many quietly told adventures. Everything was new. Everything was on a vast scale. At San
Salvador in Brazil on February 29th he notes:

> Delight is a weak term to express the feelings of a natu-
> ralist who, for the first time, has wandered by himself
> in a Brazilian forest. The elegance of the grasses, the
> novelty of the parasitical plants, the beauty of the flow-
> ers, the glossy green of the foliage, but above all the
> general luxuriance of the vegetation, filled me with ad-
> miration. A most paradoxical mixture of sound and si-
> lence pervades the shady parts of the wood. The noise
> from the insects is so loud that it may be heard even in a
> vessel anchored several hundred feet from the shore;
> yet within the recesses of the forest a universal silence
> appears to reign. To a person fond of natural history
> such a day as this brings with it a deeper pleasure than
> he can ever hope to experience again. (page 25.)

So much for his emotions, now for his speculations:

> Along the whole coast of Brazil, for a length of at least
> two thousand miles, and certainly for a considerable
> space inland, wherever solid rock occurs it belongs to a

granite formation. The circumstance of this enormous area being constituted of materials which most geologists believe to have been crystallized when heated under pressure gives rise to many curious reflections. Was this effect produced beneath the depths of a profound ocean? or did a covering of strata formerly extend over it which has since been removed? Can we believe that any power, acting for a time short of infinity, could have denuded the granite over so many thousand square leagues? (page 26.)

The answer to this problem would appear to be denudation, with the corollary of the immense antiquity of the earth.

When he got as far south as Rio he spent a few weeks in a cottage at Botofogo Bay. He was entranced. In England, he comments, any one interested in natural history who goes for a walk always has something to attract his attention; but in Brazil "the attractions are so numerous that he is scarcely able to walk at all". Let us select one. He was used to the nightingales and other singing birds of England. But nature, in climates like Brazil, "chooses her vocalists from more humble performers than in Europe. A small frog, of the genus Hyla, sits on a blade of grass about an inch above the surface of the water, and sends forth a pleasing chirp: when several are singing together they sing in harmony on different notes."

Other observations involved experimentation. He examined in Brazil a small invertebrate animal of the genus Planaria which inhabited the dry land. He was arranging his notes with such care that he was able to report at the end of his five year voyage that he had found no fewer than twelve different species of terrestrial planariae in different parts of the southern hemisphere. He tried cutting the little sluglike creature in half by a transverse section. In a fortnight the two halves had gone a long way to growing the missing parts and "in the course of twenty-five days from the operation, the more perfect half could not have been distinguished from any other specimen." "Although so well-known an experiment," Darwin concludes, "it was interesting to watch the gradual production of every essential organ out of the simple extremity of another animal." (page 43.)

A single sentence of his record of his experiences when he got to Uruguay is further proof of his astonishing zeal as a collector. "I stayed ten weeks in Maldonado, in which time a nearly perfect collection of the animals, birds and reptiles was procured."

Our next topic concerns not routine collecting but a strange and exciting discovery such as the explorer cannot hope to meet with often in his career. It exhibits also Darwin's capacity to interpret his facts and the faithfulness with which he stuck to Lyell's principle of seeking explanations of past events in observed occurrences of the present time—the principle which rescues geology from uncontrolled speculation and makes of it an observational science. Our account, while retaining Darwin's words, will be condensed from many pages of his book (pp. 105–212). Punta Alta, where the find was made, is in the Argentine.

The plain, at the distance of a few miles from the coast, belongs to the great Pampean formation, which consists in part of a reddish clay, and in part of a highly calcareous marly rock. Nearer the coast there are some plains formed from the wreck of the upper plain, and from mud, gravel, and sand thrown up by the sea during the slow elevation of the land, of which elevation we have evidence in upraised beds of recent shells, and in rounded pebbles of pumice scattered over the country. At Punta Alta we have a section of one of these later-formed little plains, which is highly interesting from the number and the extraordinary character of the remains of gigantic land-animals embedded in it. First, parts of three heads and other bones of the Megatherium. Secondly, the Megalonyx. Thirdly, the Scelidotherium, of which I obtained a nearly perfect specimen. It must have been as large as a rhinoceros. Fourthly, the Mylodon. Fifthly, another gigantic edental (i.e., without teeth in the front of its jaws) quadruped. Sixthly, a large animal with an osseous coat like an armadillo. Seventhly, an extinct kind of horse. Eighthly, a tooth of a pachydermatous animal with a long neck like a camel. Lastly, the Toxodon, perhaps one of the strangest animals ever discovered, as big as an elephant, but related to the Gnawers, the order which at the present day includes most of the smallest quadrupeds.

The remains of these nine great quadrupeds were found within the space of about two hundred yards square embedded in stratified gravel and reddish mud, just such as the sea might now wash up on a shallow bank. They were associated with twenty-three species of shells of which thirteen are recent. We may feel assured from the position in which they were found that these remains were fresh and united by their ligaments when deposited in the gravel together with the shells. Hence we have good evidence that the above enumerated quadrupeds, more different from those of the present day than the oldest of the tertiary quadrupeds of

Europe, lived while the sea was peopled with most of its present inhabitants.

While travelling through the country, I received several vivid descriptions of the effects of a late great drought. This, which occurred between 1827 and 1830, may throw some light on the cases where vast numbers of prehistoric animals of all kinds have been embedded together. During this time so little rain fell that the vegetation, even to the thistles, failed. I was informed by an eye-witness that the cattle in herds of thousands rushed into the Parana, and being exhausted by hunger were unable to crawl up the muddy banks, and thus were drowned. He added that more than once he had seen the carcasses of upwards of a thousand wild horses thus destroyed. Subsequently to the drought a very rainy season followed which caused great floods. Hence it is almost certain that some thousands of the skeletons were buried by deposits of the very next year. What would be the opinion of a geologist [Darwin means an old-fashioned, pre-Lyell geologist, still trying to save the literal truth of the story of Noah's flood] viewing such an enormous collection of bones embedded in one thick earthy mass? Would he not attribute it to a flood having swept over the surface of the land, rather than to the common order of things?

We have already quoted one of Darwin's conclusions on the basis of these observations: "Certainly, no fact in the long history of the world is so startling as the wide and repeated exterminations of its inhabitants." Another conclusion based on the same evidence comes closer to his theory of the origin of species towards which all his thoughts now tended. He observed not only that innumerable species of animals had become extinct, but that there was a close relationship between the extinct species and the living sloths, anteaters, and armadillos now characteristic of South American zoology.

This relationship is shown wonderfully—as wonderfully as between the fossil and extinct marsupial animals of Australia and those now living there—by the great collection lately brought to Europe from the caves of Brazil. In this collection there are extinct species of all the thirty-two genera, excepting four, of the terrestrial quadrupeds now inhabiting the provinces in which the caves occur; and the extinct species are more numerous than those now living. There are fossil anteaters, armadillos, tapirs, peccaries, guanacos, opossums, numerous South American gnawers and monkeys, and other animals. This wonderful relationship in the same continent between the dead and the living, will, I do not doubt, hereafter throw more light on the appearance of organic beings on our earth and their disappearance from it than any other class of facts. (page 213.)

Here we have Darwin's problem precisely defined. The new species have not simply replaced the old but have descended from them. This is what is technically called "descent with modification". The problem is to determine how the modifications arise. That was to be the subject of his *Origin of Species*.

But one thing is already clear in Darwin's mind which we may best understand by contrasting his views with those of the great French naturalist Cuvier. Cuvier

(1769–1832) had dug up from the rock formations on which Paris is built one hundred and fifty fossil species, ninety of which were extinct. Why, he asked himself, should so many whole species be destroyed? The answer he gave betrayed his Huguenot tendency to suffer a literal interpretation of the Bible to colour his scientific thinking. Where Darwin interpreted the extinction of life attested in Alta Punta as a local phenomenon due to local and natural and *still operative* causes, Cuvier read the lesson of the Paris rocks as proof of an unprecedented calamity of cosmic dimensions followed—and this is the all-important point—by a fresh divine act of creation. Darwin was aware that similar phenomena were to be found all over the surface of the globe; that to explain the advent of new species as a special act of creation in each case was a gratuitous assumption, resting on no observed evidence and explaining nothing. Moreover it ignores the fact that in each area where there is a distinct and characteristic fauna the living species retain the distinctive character of the old; the differences are enough to constitute them new species, the likeness is enough to suggest that they have descended from the old. Descent with modification must be the answer, not catastrophic destructions followed by fresh creations. The task of the scientist is to discover the law underlying the phenomena. The quest for this law was to sustain Darwin during the excitements of the rest of his voyage and during the forty years of more humdrum research that awaited him on his return.

For the moment we may still savour the excitements. When the *Beagle* had worked its passage through the dangerous Straits of Magellan and was making its slow way up the Pacific coast of South America, Darwin took the

chance to visit the island of Chiloé. From there he saw at
one and the same time three great volcanoes of the Andes
chain in eruption. It was not long before an earthquake
followed which devastated a very large part of the Chilean
coast. He made all the observations he could of its effects,
and a selection of his remarks on the psychological, social,
and geological consequences of so great a disaster will fit-
tingly be included in this chapter.

A bad earthquake at once destroys our oldest associations: the earth, the very emblem of solidity, has
moved beneath our feet like a thin crust over a fluid;
one second of time has created in the mind a strange
idea of insecurity, which hours of reflection would not
have produced. (page 365.)

Earthquakes alone are sufficient to destroy the pros-
perity of any country. If beneath England the now inert
subterranean forces should exert those powers which
most assuredly in former geological ages they have ex-
erted, how completely would the entire condition of
the country be changed! (page 369.)

It will give a better idea of the scale of these phenomena if we suppose them to have taken place at corresponding distances in Europe. Then would the land
from the North Sea to the Mediterranean have been vi-
olently shaken, and at the same instant of time a large
tract of the eastern coast of England would have been
permanently elevated, together with some outlying islands; a train of volcanoes on the coast of Holland
would have burst forth in action, and an eruption taken
place at the bottom of the sea near the northern extremity of Ireland; and lastly, the ancient vents of Au-

vergne, Cantal, and Mont d'Or would each have sent up to the sky a dark column of smoke, and have long remained in fierce action. (page 376.)

The most remarkable effect of this earthquake was the permanent elevation of the land. The land round the Bay of Concepcion was upraised two or three feet. At the island of Santa Maria the elevation was still greater; on one part were found beds of putrid mussel-shells *still adhering to the rocks* ten feet above high-water mark. The elevation of this province is particularly interesting from its having been the theatre of several other violent earthquakes, and from the vast number of sea-shells scattered over the land up to a height of certainly six hundred, and, I believe, of one thousand feet. At Valparaiso, similar shells are found at the height of thirteen hundred feet: it is hardly possible to doubt that this great elevation has been effected by successive small uprisings. (pages 374–75.)

Finally we come to Darwin's greatest contribution to geology. The voyage of the *Beagle* was now approaching its end and completing its chain of chronometrical measurements in the Pacific Ocean. On April 1, 1836, the frigate with Darwin aboard arrived at Cocos-Keeling. This was a coral island, or atoll, the only one of the hundreds he saw on which he was able to land. There are various kinds of coral reef, but an atoll is a circular ring of coral, rising only a few feet above sea level, covered with a dense foliage of coconut trees, and enclosing a lagoon whose vivid green waters contrast strongly with the deep blue of the surrounding ocean.

Cocos-Keeling lies between the Equator and the Tropic of Capricorn about 97 degrees E. In this belt of ocean, atolls abound. Their reefs, made of a limy substance, total up to an area of perhaps half a million square miles. Cocos-Keeling is six hundred miles from the nearest land, and the ocean bed at a distance of only five or six miles from the reef may be a thousand fathoms deep. It had long been recognised that the substance of these reefs is formed by the activity of what used to be called the coral insect. In Darwin's time it was already clear that insect was the wrong term. The coral zoophyte is a soft-bodied creature of tiny dimensions and of essentially the same nature as the sea anemone. From carbonate of lime, which it extracts from sea water, it makes for itself a sort of external skeleton to give it shelter and support. The little creatures themselves—the zoophytes or polyps or actinozoa—are sometimes called "soft corals" in contradistinction to their stony shelters, which are called "hard coral." They live in colonies, perhaps fifteen feet in diameter and containing a million or more polyps. The reefs are formed of the stony coral together with such ocean-borne material as may lodge on their rough surfaces.

The reefs poke up only a few feet above the surface of the sea, but may arise from depths of several thousand feet. How then are they formed? It used to be imagined that the "insects" built up their great reefs from the depths of the ocean bed. But this is quite inadmissible. The soft corals cannot exist at a depth below one hundred and fifty feet and are most active near the surface. Faced with this problem Darwin, whose imagination had been informed and disciplined by his now vast acquaintance with the behaviour of the earth's crust, and who stuck loyally to his principle of explaining past events by what could be seen going on in his own day, advanced the hypothesis that the soft corals had always, as now, built in shallow water; that the land in the Pacific area, where the coral reefs abound, had suffered a slow subsidence; and that this slow subsidence had been offset by the building activity of countless myriads of polyps. The outcome of the activity of these tiny creatures, maintained over many tens of thousands of years, had been the creation of the numerous atolls and other forms of coral reef.

The theory, as Darwin developed it, had the advantage of explaining the surprising circular form of the atolls. Imagine, before the subsidence of the ocean bed had made much progress, an island with a mountain in the middle. In the shallow waters a little offshore the polyps begin to build. As the island, through long tracts of time, slowly sinks, the encircling ring of reef is raised in height since the polyps can only live and build in shallow water. Eventually a time comes when the whole island together with its mountain has become submerged. Nothing is now to be seen but a circular reef enclosing a lagoon. On the inner side the reef slopes gradually, on the outer side, the ocean side, it plunges steeply into the depths. This is because the corals are most active in the splash and dash of the ocean waves, where there is more oxygen, more food, and more carbonate of lime to be extracted from the water. So the building goes on actively and the reef rises vertically on its outer edge. The inner bank slopes away because it initially received the detritus of the sinking island and because it still catches and holds the fragments of reef, broken off by the force of the waves, together with any other ocean-borne material cast over the reef barrier. At the same time the soil thus accumulated on the reef suffices to maintain the ring of palms that surround the lagoon.

Darwin's explanation, briefly given in *The Voyage of*

the Beagle, is worked out in full detail in his book, *The Structure and Distribution of Coral Reefs* (1842). It has met with some criticism, some reserves, some alternative suggestions as to what may have happened in special cases. But it is still generally accepted by geologists, and in the body of Darwin's work its importance is second only to the theory of the origin of species by natural selection. The mind which conceived this simple explanation of a puzzling geological problem is the same as that which conceived the theory of natural selection. We can see how the solution of the geological problem helped him in coping with the biological one which was to engage his attention for the rest of his life. For Darwin had now explained the origin of the atolls, this vast, puzzling, and beautiful feature of our world, not as an act of special creation but as the outcome of gradual, observable processes still going on before our eyes. The bewitchingly lovely multitude of coral islands and barrier reefs had been produced unconsciously and without design by the mindless life-activity of countless generations of tiny feeble creatures coping with changed circumstances in their environment.

The Scientist Comes Home

At Ascension Island on the homeward run Darwin picked up a letter from his sister. In the midst of the family news it told him how his reputation had grown in his absence. Men like the geologist Sedgwick and the botanist Henslow, a clergyman who had given him much encouragement in his natural history studies and had been the recipient of many collections sent home from the *Beagle,* were prepared to regard him as one of the leading scientists of the day. Darwin had had plenty of excitement on his trip, but also plenty of drudgery, discomfort and danger. Three of the ship's complement of officers had died of fever. Darwin himself had been bitten by a motley assortment of bugs and had been very ill. Two or three times he came near drowning. The accounts of his crossings of the Andes or scaling of rock faces in Tahiti make you hold your breath. He had not spared himself; but now he had his reward. As he started out with the letter in his pocket for the routine geological investigation in Ascension he bounded up the mountain slopes, so he tells us, like a goat.

His profession had now chosen itself. There was no more question of ordination. He made a short and very busy stay at Cambridge. He was made a Fellow of the Royal Society and returned to London where he took a post as Secretary of the Geological Association. He chose a wife, Emma Wedgwood, from a family long associated with the Darwins. She was in fact his cousin. They set up house in London and for two or three years, as well as working hard at his geological and biological notes, Darwin met all the celebrities. Then suddenly, and somewhat mysteriously, his health gave way. To avoid the strain and distraction of life in London he moved to Downe House, near Keston, in Kent. That was in 1842, when he was only thirty-three. He had still forty years of life before him, but was always an invalid. He could stroll in the country lanes but never walked the mountains again. When not in bed he was much on the sofa and always worked under difficulties. He tells us that when he was writing the *Origin of Species* he never worked longer than twenty minutes at a stretch without being interrupted by pain. Speculation about the nature of his illness is inconclusive. The doctors could then make nothing of it; but now, when more is known about tropical diseases, some are confident that he had been infected by the bite of the Bonchuca, the great black bug of the Pampas, one of the many which had dined off his blood from time to time. Others, impressed by modern knowledge of the tricks the mind can play on the body, suspect a psychosomatic illness. It was small comfort to Darwin that, in the ignorance of the nature of his disease, the suspicion of hypochondria hung about this man who had previously been so active and always remained so industrious. It is only fair to his memory to bear in mind the difficulties with which he had to contend in accomplishing his great task and to add the note of a certain dogged heroism to our estimate of his character. The voyage on the *Beagle* had taken a terrible toll of his strength. An old school-fellow who happened to meet him on his return says "he came back wasted to a shadow."

His difficulty now lay not merely in the complexity of the problem and the mass of material to be examined, but in the novelty of the solution he proposed. Everybody except for a few pioneers believed that a new species could arise only by a fresh creation. In advancing the evolutionary hypothesis, and suggesting a mechanism by which it worked, Darwin was not supported by an enthusiastic band of applauding scientists and facing the hostility only of an obscurantist Church. This is a common opinion, but the facts are against it. Let us hear what Darwin himself had to say in the sixth edition of the *Origin:*

> Although I am fully convinced of the truth of the views given in this volume . . . I by no means expect to convince experienced naturalists whose minds are stocked with a multitude of facts all viewed, during a long course of years, from a point of view directly opposite to mine. . . . A few naturalists, endowed with much flexibility of mind, and who have already begun to doubt the immutability of species, may be influenced by this volume; but I look with confidence to the future.
>
> As a record of a former state of things, I have retained in the foregoing paragraphs, and elsewhere, several sentences which imply that naturalists believe in

the separate creation of each species; and I have been
much censured for having thus expressed myself. But
undoubtedly this was the general belief when the first
edition of the present work appeared. I formerly spoke
to very many naturalists on the subject of evolution,
and never once met with any sympathetic agreement.
(pages 659–61.)

If we are to understand Darwin's new hypothesis it will
be a help to consider first the theory "directly opposite" to
his, which he tells us was that of the naturalists as well as
the theologians. It had received its fullest and clearest
statement by William Paley (1743–1805) in his *Natural
Theology, or Evidences of the Existence and Attributes of
the Deity collected from the Appearances of Nature* (1802).
This able book, which had once appealed to Darwin by
the cogency of its logic, was now judged by him, in the
light of the evidence he had collected on his travels, to be
plainly wrong.

Paley had written his book as a counterblast to older
theories of evolution such as were advanced by Lamarck
and Charles Darwin's grandfather, Erasmus. He states
their view not unfairly and rejects it.

They would persuade us to believe, he says, that the
eye, the animal to which it belongs, every other animal,
every plant, indeed every organised body which we see,
are only so many out of the possible varieties and com-
binations of being, which the lapse of infinite ages has
brought into existence; that the present world is the
relict of that variety; millions of other bodily forms and
other species having perished, being by the defect of
their constitution incapable of preservation, or of con-

tinuance by generation. Now there is no foundation whatever for this conjecture in any thing which we observe in the works of nature; no such experiments are going on at present; no such energy operates, as that which is here supposed, and which should be constantly pushing into existence new varieties of beings. (page 59.)

To Darwin, with what he now knew of the disappearance of innumerable species and their replacement by new species, this was just plain ignorance of the facts. Paley was out of date.

But there was another, more fundamental, position of Paley's which Darwin could no longer accept. The fact of the perishing of older forms and their replacement by new ones being admitted, we have still to ask how the new forms arise. It was here that the sharpest divergence showed itself. Paley insisted that the form of every existing species of plant and animal bore unmistakable witness to its creation by the divine hand. Design was everywhere visible; and most of the scientists drew from the facts the same conclusion as Paley. But Darwin had now reason to doubt this. His acquaintance with "the appearances of nature" was both firsthand and incomparably greater than Paley's, and the new evidence did not support the old views. What, in fact, did Paley maintain?

It was the famous argument from design. Nobody, said Paley, who examined a watch could fail to see design in it. There was the spring to supply the motive force, the cogged wheels to release and control it, the hands moving on the numbered dial, the glass face to render the dial visible. Given a watch we infer with confidence a watchmaker. So with the world of nature. Is the eye less obviously designed than a telescope? If from a telescope we may infer the existence and character of the instrument-maker, what must we infer from the eye? The fish has an eye adapted for seeing under water. A bird needs not only to pick up seeds but also to soar on high; its eye is marvellously adapted to both short and distant vision. "Does not this," asks Paley, "if anything can do it, bespeak an artist, master of his work, acquainted with his materials?" He has no doubt of the answer. "The most secret laws of optics must have been known to the author of such a structure."

So the argument begins. But as it develops it begins to seem faintly ridiculous. When we come to consider the internal organs, the Creator turns out to be as good at hydraulics as He is at optics. Then there is the crucial example of the lobster. For His own good reasons God put the lobster in a hard shell. "Its hardness resists expansion. How then was the growth of the lobster to be provided for? Was room to be made for it in the old shell? Or was it to be successively fitted with new ones? If a change of shell became necessary, how was the lobster to extricate himself from his present confinement? How was he to uncase his buckler, or draw his legs out of his boots?" There follows an account, at second hand, of what fishermen had reported on the process by which the lobster gets a new jacket. To clinch his point Paley concludes: "This wonderful mutation is repeated every year." (pages 238–39.)

If this now seems ridiculous to us, it did not seem so a hundred and fifty years ago. I have quoted Paley's words from the edition of 1818, and it was the eighteenth edition of a book which first appeared in 1802. It was a textbook in all the universities and seemed sound to most of Darwin's scientific contemporaries. What has made it seem ridiculous to us is the vast extension of geological and biological science in which Darwin played so great a part. The increase in knowledge exposed the utter inadequacy of Paley's conception of nature. He assumes in every case the existence of an established environment into which a new organism has to be fitted and then argues to the existence of a stupendous creator, who is not only a jack-of-all-trades but a master of all, qualified to deal with every technical and scientific difficulty as it presents itself. But the new extension of knowledge had not revealed a fixed environment into which cleverly made birds, beasts and fishes had been fitted in so skilful a manner that they had remained unaltered from the Creation to the present day. What science had revealed was a physical environment itself subject to alteration. It had revealed a mutual interdependence of environment and organism. It had revealed the destruction by changes in the environment of countless forms of life. Finally it had shown the emergence of new forms, most of them doubtless in their turn destined, too, to perish. What Darwin was looking for was the law governing the emergence of new forms. Paley's *Evidences* was no more relevant to the solution of this problem than the story of Noah's Ark.

We may now repeat and complete a quotation which expresses in his quiet way the magnitude and seriousness of the task on which Darwin felt himself embarked:

A few naturalists, endowed with much flexibility of mind, and who have already begun to doubt the immutability of species, may be influenced by this volume; but I look with confidence to the future—to young and rising naturalists, who will be able to view both sides of the question with impartiality. Whoever is led to believe that species are mutable will do good service by conscientiously expressing his convictions; for thus only can the load of prejudice by which this subject is overwhelmed be removed. (*O. S.*, page 660.)

6

The Origin of Species
by Natural Selection

At the end of 1836, when he came back from the voyage of the *Beagle,* Darwin had ceased to believe in the separate creation of new species. Two years later he lighted on the clue to his solution of this mysterious problem. Another twenty years were to go by before he published it, and even then his hand was forced. But by then he had collected such a mass of supporting evidence that his demonstration proved irresistible.

Everything that he had seen in his extensive travels had raised the problem of the origin of species in his mind. But nothing brought it into sharper focus than the visit he made, after the completion of his exploration of the South American continent, to the small Galapagos Archipelago. He returns to the subject again and again; we shall give one of his statements in his own words:

The Galapagos Archipelago lies at the distance of between 500 and 600 miles from the shores of South America. Here almost every product of the land and of the water bears the unmistakable stamp of the American continent. There are twenty-six land birds; of these, twenty-one or perhaps twenty-three are ranked as distinct species, and would commonly be assumed to have been here created; yet the close affinity of most of these birds to American species is manifest in every character, in their habits, gestures, and tones of voice. The naturalist, looking at the inhabitants of these volcanic islands in the Pacific, distant several hundred miles from the continent, feels that he is standing on American land. Why should this be so? Why should the species which are supposed to have been created in the Galapagos Archipelago, and nowhere else, bear so plainly the stamp of affinity to those created in America? There is nothing in the conditions of life, in the geological nature of the islands, in their height or climate, which closely resembles the conditions of the South American Coast: in fact there is a considerable dissimilarity in all these respects. Facts such as these admit of no sort of explanation on the ordinary view of independent creation; whereas on the view here maintained (i.e., natural selection) it is obvious that the Galapagos Islands would be likely to receive colonists from America; such colonists would be liable to modification—the principle of inheritance still betraying their original birthplace. (*O. S.*, pages 552–53, slightly condensed.)

But what precisely did Darwin mean by natural selection and how had he chanced on the idea? He had long been interested in the modification of domestic species of plants and animals by horticulturalists and breeders.

These modifications, often of the most drastic kind (as anyone who considers what has been done with roses, pigeons, dogs, and horses, to choose a few examples, will know), are effected by breeding from selected stock. But where in nature can a principle analogous to the artificial selection of the breeder be found? "In October 1838," records Darwin in his autobiography, "that is fifteen months after I had begun my systematic enquiry, I happened to read for amusement Malthus on Population, and being well prepared to appreciate the struggle for existence which everywhere goes on, from long-continued observation in the habits of animals and plants, it at once struck me that under these circumstances favourable variations would tend to be preserved and unfavourable ones to be destroyed."

T. R. Malthus, born in 1766, was a political economist who in 1798 had produced *An Essay on the Principle of Population*. The subject of this extensive and very influential survey was the relation between population and food supply. The main contention was that, while population tends to increase by geometrical progression, food production, save in exceptional circumstances, can only be advanced by arithmetical progression. Consequently there is always a pressure of population on the means of subsistence, and the growth of population is kept down by a high death-rate due to poverty, disease, war, and vice. A great proportion of births are destined not to reach maturity. "During the lapse of many thousand years there might not be a single period when the mass of people could be said to be free from distress for want of food. In every state in Europe, since we have first had accounts of it, millions and millions of human existences have been repressed from this simple cause, though perhaps in some of these states an absolute famine may never have been known."

At last Darwin had found in nature a principle of selection analogous to the artificial selection of the breeder. This is how he himself describes it in the Introduction to his book:

> I shall devote the first chapter to variation under domestication. We shall thus see that a large amount of hereditary modification is at least possible, and how great is the power of man in accumulating by his selection successive slight variations. I will then pass on to the variability of species in a state of nature. We shall discuss what circumstances are most favourable to variation. The struggle for existence amongst all organic beings throughout the world, which inevitably follows from the high geometrical ratio of their increase, will be considered. This is the doctrine of Malthus applied to the whole animal and vegetable kingdoms. As many more individuals of each species are born than can possibly survive; and as, consequently, there is a frequently recurring struggle for existence, it follows that any being, if it vary however slightly in any manner profitable to itself, under the complex and sometimes varying conditions of life, will have a better chance of surviving, and thus be *naturally selected*. From the strong principle of inheritance any selected variety will tend to propagate its new and modified form. (*O. S.*, pages 4–5.)

Having found the principle of natural selection Darwin was confident that it must in time destroy the doctrine of special creation as described by Paley. Let us have his own words again:

I am well aware that this doctrine of natural selection is open to the same objections which were first urged against Sir Charles Lyell's noble views on "the modern changes of the earth as illustrative of geology"; but we now seldom hear the agencies which we see still at work spoken of as trifling or insignificant, when used in explaining the excavation of the deepest valleys or the formation of long lines of inland cliffs. Natural selection acts only by the accumulation of small inherited modifications, each profitable to the preserved being; and as modern geology has almost banished such views as the excavation of a great valley by a single diluvial wave, so will natural selection banish the belief of the continued creation of new organic beings, or of any great and sudden modification in their structure. (*O. S.,* page 118.)

If we press the examination of the difference between Paley and Darwin a little closer, Paley's error becomes clearer still. He started off with a manufactured article, a watch, and argued from that to a watchmaker. But when we come to discuss living organisms we are not dealing with manufactured articles but with living things, and his analogy is no longer applicable. Darwin avoids direct controversy with him, but he takes up Paley's comparison of the eye with a telescope and shows how it breaks down if we remember that an eye is not manufactured but handed on from generation to generation by processes of growth and reproduction. Here are Darwin's words:

To suppose that the eye with all its inimitable contrivances for admitting different amounts of light and for the correction of spherical and chromatic aberration, could have been formed by natural selection, seems, I freely confess, absurd in the highest degree. To arrive, however, at a just conclusion regarding the formation of the eye, with all its marvellous yet not absolutely perfect characters*, it is indispensable that the reason should conquer the imagination; but I have felt the difficulty far too keenly to be surprised at others hesitating to extend the principle of natural selection to so startling a length.

It is scarcely possible to avoid comparing the eye with a telescope. We know that this instrument has been perfected by the long-continued efforts of the highest human intellects; and we naturally infer that the eye has been formed by a somewhat analogous process. But may not this inference be presumptuous? Have we any right to assume that the Creator works by intellectual powers like those of man? If we must compare the eye to an optical instrument, we ought in imagination to take a thick layer of transparent tissue, with spaces filled with fluid, and with a nerve sensitive to light beneath, and then suppose every part of this layer to be continually changing slowly in density, so as to separate into layers of different densities and thicknesses, placed at different distances from each other, and with the surfaces of each layer slowly changing in form. Further we must suppose that there is a power, represented by natural selection or the survival of the fittest, always intently watching each slight alteration in the transparent layers; and carefully preserving each which, under varied circumstances, in any way or in any degree, tends to produce a distincter image. We must suppose each new state of the instrument to be

---

*Darwin uses "characters" where we say "characteristics."

multiplied by the million; each to be preserved until a better one is produced, and then the old ones to be all destroyed. In living bodies, variation will cause the slight alterations, generation will multiply them almost infinitely, and natural selection will pick out with unerring skill each improvement. Let this process go on for millions of years; and during each year on millions of individuals of many kinds; and may we not believe that a living optical instrument might then be formed as superior to one of glass as the works of the Creator are to those of man? (*O. S.*, pages 223–28 condensed.)

Here Paley's outmoded conception of the universe as a complex mechanism produced by a super-craftsman is replaced, and replaced forever, by the vastly more informed conception of an evolving universe. But we note also that, while the notion of special creations is dropped, the idea of a Creator remains. Indeed the passage can be read, or rather must be read, not only as a protest against Paley's inadequate view of the creation but equally against his inadequate view of God. The lesson of the closing pages of the book is the same:

Authors of the highest eminence seem to be fully satisfied with the view that each species has been independently created. To my mind it accords better with what we know of the laws impressed on matter by the Creator, that the production and extinction of the past and present inhabitants of the world should have been due to secondary causes, like those determining the birth and death of the individual. When I view all beings not as special creations, but as the lineal descendants of some few beings which lived long before the first bed of the Cambrian system was deposited, they seem to me to become ennobled. (*O. S.,* page 668.)

If again we ask Darwin what are those laws impressed on matter, those secondary causes by which the Creator prefers to operate, this is his answer:

These laws, taken in the largest sense, are Growth with Reproduction; Inheritance which is almost implied by reproduction; Variability from the indirect and direct action of the conditions of life and from use and disuse; a Ratio of Increase so high as to lead to a Struggle for Life, and as a consequence to Natural Selection, entailing Divergence of Character and the Extinction of less-improved forms. Thus, from the war of nature, from famine and death, the most exalted object which we are capable of conceiving, namely, the production of the higher animals, directly follows. There is grandeur in this view of life, with its several powers, having been originally breathed by the Creator into a few forms or into one; and that, while this planet has gone cycling on according to the fixed law of gravity, from so simple a beginning endless forms most beautiful and most wonderful have been, and are being, evolved. (*O. S.,* pages 669–70.)

So ends the epoch-making *Origin of Species.*

The Descent of Man

In reading *Origin of Species* no one doubts that in his own mind Darwin includes man among the products of natural selection. But he does not say so. Once only he alludes to the topic in a gingerly way and then drops it like a hot potato. "Much light will be thrown on the origin of man and his history." (page 688.) That is all. And from the eloquent peroration, quoted at the end of our last chapter, man disappears. There, surprisingly, it is the higher animals, not man, who are called "the most exalted object we are capable of conceiving." Even for a nation of horse-and-dog-lovers this is going rather far. Hamlet calls man "the paragon of animals." In the Psalms man is only "a little lower than the angels." But with Darwin the higher animals step up into the first place. Nobody writes like this without a cause. The fact is that Darwin still shrank from presenting man as a product of the blind force of natural selection.

This caution, or hesitancy, is reflected in the dates of Darwin's publications. "In July 1837," he tells us in his *Autobiography,* "I opened my first notebook for facts in relation to the *Origin of Species*." By 1844 he had completed his statement of the case in an immensely long Essay. This he did not publish, but showed to his friend Dr. Hooker, in order, we must suppose, to establish his claim to priority. Then fourteen years later, in June 1858, Alfred Russell Wallace, who had independently arrived at the same theory, sent him a short paper on it to be presented, if Darwin thought fit, to the Linnaean Society. Thus the question of publication was at last brought to a head. Darwin took council with his friends and the satisfactory compromise was agreed that Darwin and Wallace should jointly present papers to the Linnaean Society, in July 1858. This, the first announcement of the theory of Natural Selection, produced little effect. But Darwin, whose priority was clear, and whose statement was backed by a mass of evidence unknown to Wallace, at last took the matter of publication resolutely in hand. He greatly reduced the amount of material he had incorporated in the Essay of 1844 (so much so that he calls the finished book merely an "abstract" of his material) and the first edition of the *Origin of Species* came from the press in November 1859. This time the impact of the theory was immense. Some of those best qualified to judge were, understandably enough, the most reluctant to give in. Bishop Wilberforce gets the blame for leading the attack on it, but it was a brother scientist, the palaeontologist Owen, who supplied the bishop with his facts. But the date is important as marking the moment when the general reading public became interested in the theory, ready to discuss it, and favourably impressed. The year 1859 is rightly taken as a landmark in intellectual history. Both the theologians and the scientists have been digesting Darwin's findings ever since.

Having broken the ice with the *Origin of Species* Darwin was bound sooner or later to follow with a pronouncement on the origin of man. He, in fact, waited another twelve years before, in 1871, the two volumes of *The Descent of Man* were given to the public. When they did appear they created much less opposition than the earlier work had done, partly, no doubt, because many readers of the *Origin* had made the obvious deduction for themselves and also because a host of lesser writers had done it for them. The public were already prepared to accept the anthropoid apes as members of the family. Darwin's new work could only add the authority of the great man to a familiar opinion and draw out the wider implications of the place of man in the animal kingdom. It is Darwin's understanding of these wider implications that is illustrated in the quotations that follow.

Chapter One is called *Evidence of the Descent of Man from some lower Form*. Here is a selection of the evidences:

It is notorious that man is constructed on the same general type or model with the other mammals. All the bones in his skeleton can be compared with the corresponding bones in a monkey, bat, or seal. So it is with his muscles, nerves, blood-vessels, and internal viscera. The brain, the most important of all the organs, follows the same law. (page 10.)

Man is liable to receive from the lower animals, and to communicate to them, certain diseases. This fact proves the close similarity of their tissues and blood, both in minute structure and composition, far more plainly than does their comparison under the microscope, or by the aid of the best chemical analysis. (page 11.)

Medicines produce the same effect on them as on us. Many kinds of monkey have a strong taste for tea, coffee, and spirituous liquors; they will also, as I have myself seen, smoke tobacco with pleasure. These trifling facts prove how similar the nerves of taste must be in monkeys and man, and how similarly their whole nervous system is affected. (page 12.)

It is, in short, scarcely possible to exaggerate the close correspondence in general structure, in the minute structure of tissues, in chemical composition and in constitution, between man and the higher animals, especially the anthropomorphous apes. (page 14.)

Man and all other vertebrate animals have been constructed on the same general model, pass through the same early stages of development, and retain certain rudiments in common. Consequently we ought frankly to admit their community of descent. It is only our natural prejudice and that arrogance which made our forefathers declare that they were descended from demigods, which lead us to demur to this conclusion. But the time will before long come when it will be thought wonderful that naturalists, who were well acquainted with the comparative structure and development of man and other mammals, should have believed that each was the work of a separate act of creation. (pages 32–33.)

Chapter One was concerned only with the physical structure and physiological behaviour of men and animals. Chapter Two carries the comparison a stage further. It is entitled *Mental Powers*. In it we read:

> If no organic being excepting man had possessed any mental power, or if his powers had been of a wholly different nature from those of lower animals, then we should never have been able to convince ourselves that our high faculties had been gradually developed. But it can be clearly shown that there is no fundamental difference of this kind.
>
> My object in this chapter is to show that there is no fundamental difference between man and the higher mammals in their mental faculties. (page 35.)

> There is no evidence that man was aboriginally endowed with the ennobling belief in the existence of an Omnipotent God. (page 65.)

Chapter Three goes still further. Its title is *Moral Sense,* and its purpose is to bring the whole moral, as well as the physical and mental, life of man under the control of natural selection:

> I fully subscribe to the judgment of those writers who maintain that of all the differences between men and the lower animals, the moral sense or conscience is the most important. (page 70.)

> The following proposition seems to me in a high degree probable—namely, that any animal whatever, endowed with well-marked social instincts, would inevitably acquire a moral sense or conscience, as soon as

its intellectual powers had become as well developed, or nearly as well developed, as in man. (page 71.)

The difference in mind between men and the higher animals, great as it is, is certainly one of degree and not of kind. The senses and intuitions, the various emotions and faculties, such as love, memory, attention, curiosity, imitation, reason, etc., of which man boasts, may be found in an incipient, or even sometimes in a well-developed condition, in the lower animals. (page 105.)

Chapter Four, *The Manner of Development,* carries the argument a stage further still. The object now is to show the law of natural selection operative in the social as well as the biological development of man. As the same argument is also the theme of Chapter Five, *The Civilised Nations,* and Chapter Six, *On the Affinities and Genealogy of Man,* we shall run our extracts from these three chapters on continuously, indicating only the pages where they may be found:

It is manifest that man is now subject to much variability. No two individuals of the same race are quite alike. We may compare millions of faces and each will be distinct. There is an equally great amount of diversity in the proportions and dimensions of the various parts of the body. The variability or diversity of the mental faculties in men of the same race, not to mention the greater differences between men of distinct races, is so notorious that not a word need here be said. So is it with the lower animals. (pages 108–9.)

We have now seen that man is variable in body and mind; and that the variations are induced, either directly or indirectly, by the same general causes, and obey the same general laws, as with the lower animals. Man has spread widely over the face of the earth, and must have been exposed, during his incessant migrations, to the most diversified conditions. The inhabitants of Tierra del Fuego, the Cape of Good Hope, and Tasmania in the one hemisphere, and of the Arctic regions in the other, must have passed through many climates and changed their habits many times, before they reached their present homes. The early progenitors of man must also have tended, like all other animals, to have increased beyond their means of subsistence; they must occasionally have been exposed to the struggle for existence, and consequently to the rigid law of natural selection. Beneficial variations of all kinds will thus, either occasionally or habitually, have been preserved and injurious ones eliminated. (page 135.)

Man in the rudest state in which he now exists is the most dominant animal that has ever appeared on earth. He has spread more widely than any other highly organised form; and all others have yielded before him. He manifestly owes his immense superiority to his intellectual faculties, his social habits, which lead him to aid and defend his fellows, and to his corporeal structure. The supreme importance of these characters has been proved by the final arbitrament of the battle for life. (page 136.)

He who was ready to sacrifice his life, as many a savage has been, rather than betray his comrades, would often leave no offspring to inherit his noble nature. But a tribe including many members who were always ready to give aid to each other and to sacrifice themselves for the common good, would be victorious over most other tribes; and this would be natural selection. At all times throughout the world tribes have supplanted other tribes; and as morality is one element in their success, the standard of morality and the number of well-endowed men will thus everywhere tend to rise and increase. (pages 163 and 166.)

To believe that man was aboriginally civilized, and then suffered utter degradation in so many regions, is to take a pitiably low view of human nature. It is apparently a truer and more cheerful view that progress has been much more general than retrogression; that man has risen, though by slow and interrupted steps, from a lowly condition to the highest standard as yet attained by him in knowledge, morals, and religion. (pages 183–4.)

We have given to man a pedigree of prodigious length, but not, it may be said, of noble quality. The world, it has often been remarked, appears as if it had long been preparing for the advent of man; and this, in one sense, is strictly true, for he owes his birth to a long line of progenitors. If any single link in this chain had never existed, man would not have been exactly what he now is. Unless we wilfully close our eyes, we may, with our present knowledge, approximately recognise our parentage; nor need we feel ashamed of it. The most humble organism is something much higher than the inorganic dust under our feet; and no one with an unbiased mind can study any living creature, however humble, without being struck with enthusiasm at its marvellous structure and properties. (page 213.)

We have now summarised in his own words Darwin's account of the evolutionary process as it applies, first, to all species and, then, to man. In arriving at his interpretation of the development of life at all stages Darwin rested on a vast, orderly collection of evidence as to the facts of natural history such as in its entirety had probably never before been held together in any one mind. In collecting the evidence he was borne up by a passionate zeal matched only by the dogged patience with which he applied himself to its interpretation. Moreover his candour in expressing the objections to his theory and even admitting his doubts are, together with his other qualities, an example and inspiration to all scientists.

But no wonder that his findings were also felt to be disconcerting. We learn that natural selection, the main principle of evolution and Darwin's special contribution to the theory, is an entirely blind and automatic process pervading the whole organic world yet working unfailingly for the progress of such species as survive. We notice also that in Darwin's mind it governs the evolution not only of the lower animals but of man. Nor does he limit its action to man as a biological organism. Natural selection is also held to be the clue to man's progress mentally, morally, and in all the refinements of civilization including religion. It is through the blind and aimless power of natural selection that man has advanced to his present level in "love, memory, attention, curiosity, imitation, reason, etc." It is through natural selection that he has progressed in "knowledge, morals, and religion" and so acquired the willingness to "sacrifice himself for the common good" together with "the ennobling belief in the existence of an Omnipotent God."

Darwin and His Predecessors

Evolution had had a history of about a hundred years before Darwin announced the theory of natural selection. It is characteristic of Darwin that in writing his *Origin of Species* he makes no mention of the contribution of his predecessors. He writes as if his experiences on the voyage of the *Beagle* had put the idea into his head and his protracted researches of the next twenty years or so had confirmed and clarified his views. Only in the third edition of his book, after many thousand copies had been sold, did it occur to him to prefix a short, and not very informative "Historical Sketch of the progress of opinion on the origin of species previous to the publication of the first edition of this work." This is all the more surprising when we reflect that one of the greatest of the pioneers was his own grandfather. The explanation, I think, is simple. As I wrote in the first chapter, he was a naturalist of the first order but had little interest in books, that is to say, in the history of ideas.

Three of his predecessors will be mentioned here. The great French naturalist, Buffon (1707–1788), published in 1749 his *Theory of the Earth*. In it he sets aside the then current practice of basing natural history on the interpretation of the Scriptures. He declares his opinion that the earth is very old, being in all probability a fragment of the sun detached by the glancing blow of a comet. Its various changes from its origin to its present state he seeks to interpret, like Lyell after him, in the light of observable processes still going on. He followed this up in 1778 by his *Epochs of Nature* in which, still sticking to his principle of gradual change by observable causes, he attempts to fix the chronological order of the appearance on earth of different species. But as a clue to these changes he had only the direct effect of the environment to suggest. Still the fact remains that he was teaching the doctrine of descent with modification and that the popularity of his many books throughout Europe was immense.

Of Erasmus Darwin (1751–1802) we have already spoken. He, too, believed in descent with modification and also stressed the inherent activity of the organism as the main means of evolution. It remains to add that his reputation as a writer both in verse and prose was so great that among contemporaries in Europe he was overshadowed only by Goethe.

In the third place Lamarck (1744–1829) developed the theory of an inherent faculty of self-improvement by his teaching that new organs arise from new needs, that new organs continue to develop in proportion to the extent to which they are used, and that these acquisitions are handed down from one generation to the next. Conversely, disuse of existing organs leads to their gradual disappearance.

None of this long history of evolutionary theory was unknown to Charles Darwin and none of it did he ever totally discard. His contribution was to add to it, as the principle means of evolution, the theory of Natural Selection. It is a thousand pities that he did not preface the first edition of his book with an historical sketch, for it would have greatly facilitated the understanding of his own contribution. But it is probable that this kind of exercise was outside the range of his mental powers. Otherwise it is hard to explain the inadequacy of his sketch when he did get down to it. Erasmus Darwin, for example, is mentioned only in a footnote and then almost with contempt. "It is curious how largely my grandfather, Dr. Erasmus Darwin, anticipated the views and erroneous grounds of opinion of Lamarck."

To sum up: Charles Darwin inherited from the older evolutionists the doctrine of descent with modification; the doctrine of the direct or indirect action of the environment on the individual organism; and a fluctuating belief in the Lamarckian principle that new characteristics acquired by the individual by use or disuse are handed on to its descendants. To this he added the idea that the main cause of modification is variations that spontaneously arise in the passage from one generation to the next. The cause of these variations is unknown, but if they happen to be useful they survive by natural selection. Natural selection operates because organisms tend to multiply by geometrical progression while the means of subsistence cannot do so. Consequently there is a weeding out in each generation. The least fit perish, and any random variation which promotes the chance of survival means that the possessor of the favourable variation will live to hand it on to the next generation. The accumulation of favourable variations over long periods of time will result in the emergence of a new species and in the extinction of the older and less well-adapted species. Thus, though the action of the principle of natural selection is blind, it is nevertheless a law of progress and its operation extends over the whole organic realm from the amoeba to man.

In our next chapters we shall consider how this theory has stood the test of time.

The Rise of Genetics

Genetics is concerned with the mechanism of inheritance. Though the problem of inheritance was Darwin's great concern, the mechanism of it completely escaped him. It is true, of course, that Darwin had focused attention on reproduction as the source of the variations on which natural selection worked. He was not concerned with the differences between adults of the same species, but with the differences between one generation and the next. In general, species breed true, but there are occasional small variations. Inheritance and variation were facts of nature on which Darwin had amassed more information than any other man. His information had sufficed to enable him to establish the principle of natural selection as the main cause of the origin of new species. But the mechanism of inheritance was a nut he could not crack.

The advance, when it did come, turned on the study of the cell. Darwin knew that plant and animal tissues are cellular. He also knew that cells arise from existing cells, not spontaneously from inorganic matter. Virchow in 1855 had established the proposition *omnis cellula e cellula,* every cell from a cell. Virchow knew that the female ovum was a cell, and in 1865 it was shown that the male spermatozoa are cells too. The reproductive cells are the link between the generations. If inheritance is to be understood it is the reproductive cells that must be studied. In the 1880s this was strongly emphasised by Weismann. He drew a sharp distinction between the soma (i.e., the body) and the germ-plasm (i.e., the reproductive cells). It is the union of two cells, one male and the other female, that initiates the growth of a new individual; but the reproductive cells in the parents are the direct source of the reproductive cells in the children. The soma or body is the creation of the fertilized ovum. But the reproductive cells, the ovum and spermatozoa, are not the product of the body as a whole but of the germ-plasm. The successive generations of perishable individuals are only a sort of shelter created by the germ-plasm for itself, which, in contrast to the body, enjoys a sort of immortality. This theory explains the fallacy in the view of Lamarck, who believed that acquired characteristics were heritable. Acquired characteristics, like the smith's brawny arms or the pianist's nimble fingers, are attributes of the soma and cannot be passed on to the next generation.

Weismann had taken the first step towards answering Darwin's problem. Another great step, unknown to Darwin or to any other scientist of the time, was taken by an Austrian Abbé, Gregor Mendel, who, like many other horticulturalists before him, had interested himself in the growing of peas. Mendel had read *Origin of Species* and had acquired from it an interest in evolutionary theory. He instituted a prolonged course of experimentation in the propagation of peas and, in 1866, a paper he read, to the Natural History Society of the little town of Brunn, was printed locally. It attracted no attention, but it is interesting to speculate what effect it would have had on Darwin's work if he had read it, for Darwin had still sixteen years to live during which he maintained a line of speculation which Mendel had rendered out of date.

Mendel's experiments exhibit the simplicity of genius. His purpose was, by his experiments in breeding peas, to throw light on the laws of inheritance. He selected for examination seven characteristics in which peas differ. These concern points like seed shape, seed colour, pod shape, pod colour, and so on. The differences are that some peas are round and some wrinkled, some yellow and some green, and so on. In each experiment he proposed to enquire into the laws of inheritance of one quality, by crossing round peas with wrinkled, or yellow peas with green, and so on. For brevity we shall confine ourselves to the question of shape. He crossed round peas with wrinkled and awaited the result. This was that all the next generation of peas were round. And so it was with the other six pairs of characters tested. The result of the crossing was never a mixture of characteristics. The new generation of peas was like one or other of the parents, not a blending of both. To use the now established terminology, one quality was found to be dominant and the other recessive.

Mendel now had another question to put to his peas. To the sight there was no difference between the parent round peas and those of the next generation. Both were round. But Mendel knew that the parent round peas were descended from two round parents, while the next generation of round peas had one round parent and one wrinkled. Their genetic makeup was different though it did not show. He accordingly bred a second generation of peas from round peas of mixed parentage. The result confirmed his suspicions. In the second generation 75 percent of the progeny were round, but 25 percent had reverted to the wrinkled shape. In all the experiments, whatever quality was being tested, the results were the same—in the second generation in 25 percent of cases the recessive character reappeared.

Mendel carried his experiments into a third generation, but we need not follow him there. The results he had established suffice. First he had drawn a clear distinction between the appearance of an individual (its phenotype) and its genetic composition (its genotype). Next he had revealed that the inheritance of characters is atomic or particulate. In other words, the inherited quality is not a blend of those of the two parents but either one or the other. The quality in the offspring is the result of some "factor," as Mendel called it, in the parents. These "factors" are now called genes.

It was not till 1900, when the improvement of the microscope itself and the technique of its use had made much progress, that the internal structure of the cell began to be visible; then, as a result of renewed interest in genetics, Mendel's paper was discovered independently by three researchers in different countries, who were combing through the literature to acquaint themselves with what had already been done. This was really the birth of modern genetics. What Mendel had seen with the naked eye more than thirty years earlier now began to find its explanation in what microscopy revealed. This is what came to light.

Every cell in the body has a nucleus. The nucleus contains a fixed number of rodlike, or threadlike, "chromosomes," characteristic of the species concerned. (Chromosome means coloured body. The name is derived from the new technique of staining which made them visible under the microscope.) All cells multiply by division. The genes, the real carriers of the heritable material, are strung along the length of the chromosomes; and they are strung in pairs. Only in the germ-cells, as distinct from the body-cells, is this not so. In them out of each pair of genes only one gene is present. Fertilization occurs when a male cell (spermatozoon) penetrates a female cell (ovum). Then there is a fusion of the chromosomes, so that the fertilised egg has a full complement of genes made up from both parents. The existence of genes was inferred from their effects long before they could be seen. They did not become visible till the electron microscope had been developed.

Meanwhile the chemists, from their angle, were tackling the problem of the chemical composition of genes. The solution of this problem was a major step forward, bringing a crop of Nobel Prizes in the 1960s. It appears that the basic chemical constituent of the gene is the immensely complex molecule DNA (short for deoxyribonucleic acid). This giant molecule, which is yet only of the scale of magnitude of a virus, is the chemical bearer of the material on which inheritance depends. These infinitesimally small entities control both the structure and the functioning of the cell. In that very special cell, the fertilized ovum, it is the genes that issue instructions for the building of a new individual, whether a mouse or a man. It is said that, if the amount of "information" requisite for this purpose were expressed in human speech, it would need an encyclopaedia more voluminous than any yet compiled on earth.

This new discovery made it clear that Mendel was right in his theory of particulate inheritance. It reinforced Weismann's distinction between the soma and the germ-plasm and thus ruled out the Lamarckian notion of the inheritance of acquired characters, that is characters acquired by the soma in the course of its life. For the structure of the DNA molecule is such that, while it can issue instructions for the making of a new individual, the new individual, when made, has no means of sending information back to the germ-plasm. Accordingly acquired bodily characteristics, like the distant sight of the mountaineer or the near sight of the compositor, are not passed on to the children. Lamarck was wrong to think the giraffe got its long neck by straining generation after generation to reach the foliage on taller trees.

It remains to consider the all-important question of the significance of the new genetics for the problem of variation. Where, in the light of the new knowledge, is the source of the variations between one generation and the next on which Darwinian natural selection acts? The effect is not to deprive natural selection of its fund of variations, but to explain more clearly how they arise. There are two main sources. First, there is the reshuffling of the genes which takes place at fertilisation. It is now recognised that genes are not necessarily related to somatic "characters" in any one-to-one way. In the reproduction of the immense variety of "characters" discernible in any individual being, a single gene plays many roles and does so in combination with other genes. So it comes about that the reshuffling of the gene-pack at fertilisation leaves room for a new hand to be dealt.

Of equal, or probably of greater, importance is the discovery of "mutation" in the genes. Normally the stability of genes is such that a gene will reproduce itself through a hundred thousand generations without any change. Then suddenly something changes. The gene becomes a "mutant gene" and issues different instructions for the building of that part in the new individual for which it is responsible. These rare, and random, mutations provide the fund of variation on which natural selection works. The doctrine of natural selection, as Darwin defined it, is not superseded but rendered more precise and given a better base in observed fact.

But modern evolutionists do not describe the process in Darwin's rosy terms. Mutations, though they occur at calculable intervals of time, are random and blind. They do not automatically present nature with improvements to adopt. "Most evolutionary change has been degenerative," said J. B. S. Haldane bluntly. "Natural selection does not guarantee progress," says Julian Huxley. Darwin himself, of course, taught that the emergence of every new species is balanced by the extinction of another less well adapted to survive. He insisted that there was no more reason to boggle at the extinction of a species than of an individual. But contemporary evolutionists make a different point. They stress that evolutionary history is not only cruel, blind, and mechanical, but also that it is full of dead ends. No progress has occurred in bees or ants since the oligocene period, nor in birds since the miocene. None in mammals since the pliocene period, and then only in man. And as for man, his progress has been achieved by means quite other than those with which Darwin busied himself.

But here we come to a development in evolutionary theory greater even than the rise of classical genetics and even less within Darwin's ken. We reserve it for the next chapter.

The Uniqueness of Man

In his *Origin of Species*, in deference to the prejudices of the age, Darwin avoided mention of man. In his *Descent of Man* he goes to the opposite extreme. He not only maintains the reasonable proposition that in physical structure and physiological behaviour there is no fundamental distinction between man and other mammals. He goes on to argue that the same proposition holds true with regard to man's mental and moral powers. This has proved an unsound speculation and serves to indicate Darwin's limitations as a thinker. We may recall the judgment of Dr. Charles Singer, an excellent historian of science. "Darwin," he writes, "was an investigator of the very first rank, but he was inexpert in the exact use of language, and had little philosophical insight." This is a just judgment, and it is well to remember that knowledge cannot be advanced merely by observation. The most brilliant observer still needs to have a mental grasp of the subject of his investigations. This Darwin had in a unique degree in the geological and biological sphere: he was a superb naturalist. But it deserted him in the human sphere. He was a poor philosopher.

We have already quoted his opinion that "any animal whatever, endowed with well-marked social instincts, would acquire a moral sense or conscience, so soon as its intellectual powers had become as well developed, or nearly as well developed, as in man". This is only a roundabout way of saying that any animal that became, or nearly became, a man would have a conscience. But what other animal is on the way to develop intellectual powers equal to those of a man? And how does a social *instinct* become transformed into a social *conscience?* Instincts are biologically inherited patterns of behaviour carried out automatically without conscious purpose. Morality is something quite different. So far as ethnological research has been able to push its enquiries back into the simplest forms of human society, men are always found to be governed by complex codes of behaviour. But these rules are not instinctive. They are not biologically inherited. They are institutions whose origins are referred to remote ancestors; they are inculcated by education; they involve the use of language. In short, they belong to the realm, not of instinct, but of mind. They are human facts, not animal facts.

Nowhere in Darwin's writings can we find proof that he grasped this distinction. Hence his constant endeavour to bring the life of the mind under the laws of biological evolution. There are some crude speculations in his Notebooks which have a bearing on this failure. "Why," he asks himself, "is thought being a secretion of the brain more wonderful than gravity a property of matter?" He recognizes the crude materialism of this notion but rather likes it. He expostulates playfully with himself: "Oh, you materialist!" Then, after more of the same kind, he concludes: "Now that I have a taste of hardness of thought. . . ." He is taking stock of his progress and thinks he is getting on.

But what, really, are we to make of his definition of thought as a secretion? On the face of it, it is a failure to distinguish the physiological from the psychological. Suppose we say: "A stitch in time saves nine" or "A rolling stone gathers no moss." Here we have a pair of homely thoughts to use as a touchstone of the new definition. Simple as they are they involve wide generalisations arising out of experience of human society. They tell us in a symbolic way that when things begin to go wrong they can easily be set right if tackled promptly. Or that a man who shifts from one job to another will not end up rich. But simple and homely as these samples of proverbial wisdom are, they depend on conceptual thinking and involve the use of words. What, then, is the sense of calling them a secretion? Language is an essential difference between animals and men. What then is the sense of defining language in terms of the physiology common to men and animals? Darwin might have asked himself why the brain, being an organ common to all mankind, secreted Hebrew in Jerusalem, Greek in Athens, and Latin in Rome. He might have, but he didn't. The truth is that Darwin never came in sight of the distinction between a biological fact and a mental fact.

In modern evolutionary theory this confusion has been overcome. In evolution as applied to man two stages are recognised—the animal stage and the distinctively human stage. The second, the psycho-social stage, considers man in society. In this stage the development of man no longer depends on the blind action of natural selection but is guided more and more by the purposive action of man himself. Man is the sole bearer of evolution at this stage, and in this his uniqueness consists.

Evolution at the
Psycho-Social Stage

Man is unique in being the only tool-making animal and the only talking animal. He is in consequence uniquely educable. He still, of course, transmits his physical inheritance of genes from generation to generation; but his mental life, his culture, is passed on by training, educationally not biologically.

These distinctive characteristics of man have made his relation to the natural environment different from that of any other animal. In biological evolution the pressure of the environment is beyond the control of the animal populations subject to it. Not so with man. Since man has been man he has been purposively engaged in changing his natural environment. The control of fire, clothing, shelter, food-production and food-storage, have made it possible for man to survive in almost any climate. Every step in advance has meant the adaptation of the natural environment to man, not of man to the natural environment.

It is customary now to consider the life of any organism not solely in itself, but in its natural setting. This shift of emphasis has given rise to the all-embracing science of ecology, that is, the study of the habitat. From this angle also the uniqueness of man is revealed. His environment is largely of his own making. His habitat is not any longer nature in the raw but nature adapted by man to his own ends. Human ecology is unique.

When man learned to control fire; when he domesticated the dog (or perhaps we should say, when the dog adopted him); when he learned agriculture and horticulture, and domesticated the sheep and the ox; every step meant a humanising of the environment. Equally true is this of his discovery of the art of navigation and his various engineering feats on land. All these advances by which, in the Biblical phrase, he came to replenish the earth and subdue it, were so many steps towards making the earth his home. As a contemporary poet puts it:

> Home was not built in a day, nor by the beasts and
>   birds.
> The robin keeps not open nest, the burning tiger
> For all his jungle-craft invites no friend to dine
> And share the modest comforts of the ancestral lair.
> Whence man, beginning as a beast among beasts
> (Only his snobbery would disallow bloodkinship),
> Is driven at last to break with his poor relations,
> Cast off the brute, bring out the man, and incarnate
> The humane animal, before he can set up house.
> —Leonard Barnes, *The Homecoming*

Now, of all the tools man has acquired to break with his poor relations and set up house in his planet, the most important one is speech. Without this all else would be impossible. There could be no society, no organised endeavour, no concerted purpose. We must, therefore, think a little about the nature of speech. Some zoologists, in seeking to throw light on the origin of speech, have concentrated attention on the vocal sounds of beasts and birds. This is only another example of the folly of accumulating observations without first acquiring a mental grasp of the subject under investigation. The essential character of language is not sound but the communication of information, and that by means of symbols. The symbols may be visual or auditory. The communication may be made as well by signs, gestures, dances, or drawings as by words. But the signs, whatever they are, must be agreed symbols for the things to which they refer. There is no intrinsic connection between a symbol and the thing symbolised. The connection has to be established as a convention and subsequently learned.

How man acquired this power of speech depending on the use of conventional symbols is a mystery not fully explored. No doubt it was a slow acquisition, and if we could follow the steps by which it was accomplished, it would seem less mysterious. Here we can do no more than define the question a little more closely. It is now generally agreed that man, the tool-maker, *homo faber,* has been on earth a million, if not two million, years. But it is very probable that it was not till about the date of the Lascaux cave drawings, say about thirty thousand years ago, that man became fully articulate. In other words, *homo sapiens* is less than fifty thousand years old; and hundreds of thousands of years before that went to the perfecting of his greatest invention, speech. But it is not difficult to see how the essential characteristic of speech is implicit in the making of tools. Speech is the communication of information through visible or auditory symbols. A man who makes anything must first have the idea of it in his mind. The dissemination of the art of making anything, even the simplest stone tool, is the dissemination of information, the passing from one man to another of the idea of the thing to be made and of the various processes of its manufacture. No doubt at first the instruction was better conveyed by actions than by spoken words. Gestures are older symbols than words; and vocal sounds, no doubt, were at first accompaniments of the instruction rather than its most important element. But vocal symbols have great advantages. They can be heard in the dark; they leave the hands free; and as soon as tongue, teeth and lips become more practised, the voice can produce an immensely greater variety of clearly distinguishable symbols than the hands. So speech made its way and rational discourse became the chief characteristic distinguishing man from the other creatures.

Because speech is only a symbol, a contrast has often been drawn between things, as real, and words, as mere shadows of things. This is true up to a point. The word "bread" provides no nourishment. But, all the same, words can do what things cannot do. They are, in fact, a special kind of thing. The difference is this. Things act directly on our senses and satisfy our physical needs. Words are addressed to the mind. The stimulation of our senses and the satisfaction of our needs are individual, private matters. If we had not speech we would be wrapped up in our physical selves. Words are not only the means by which we communicate with one another; without words consciousness itself is dim. Words are immaterial, mental things; but without words there would be no control of things. Words release us from bondage to our immediate sensations. Without words there could be no society, no purposive cooperation, no creation of a new environment for ourselves. They are not, then, shadows of things but the very substance of our new distinctive human life. They are both individual and social. It is I who speak, but in speaking I communicate with you, and employ a medium I did not create—a medium which has no meaning whatsoever except insofar as I am a part of the society which created it. Language makes a man conscious of himself, but conscious of himself as a member of society.

Hence, when we leave the animal sphere and enter the human, we leave the realm of biology behind and enter the psycho-social realm. In the psycho-social realm, and more particularly since *homo faber* became *homo sapiens,* man gradually emancipates himself from the law of natural selection. His evolution ceases to be a biological process and becomes a conscious one. The incredibly slow, cruel, blind and blundering process of natural selection is superseded. Man progresses through his own inventions; and these are passed on from one generation to the next by education and accumulate with a rapidity without parallel in the biological phase. Also the pace of change accelerates. More is accomplished in one generation by *homo sapiens* than in a hundred by *homo faber.*

Nor is it only the speed of evolution that changes. Its character also is new. When man was still in the animal stage he was under the iron law of determinism. He was pushed on from behind in accordance with the law of natural selection. But since man became man the changes in his way of life are the consequences of his conscious actions. It is possible for him to have a vision of a goal at which he aims. Hence, in comparison with the biological stage, the psycho-social stage appears as a period of freedom and creativity. The art of making stone tools did not *happen* to man, like the biological development of a new organ; it was something that he *did*. So with speech, so with writing; so with the domestication of plants and animals; so with the arts and crafts; so with the creation of institutions like family and clan, like tribe and city, like nation and state. So with religions, codes of law, literature, music, architecture and painting. These are all achievements of man, monuments of his invention, solutions of his problems, tributes to his creativity.

Hence in the psycho-social stage it is creativity we should stress. The process by which something new comes into existence is somewhat of a mystery. There is an element of the incalculable and the unforeseeable about it, and older generations of scientists were apt for that reason to ignore it. It is natural for the tidy mind to bring things under the rule of law. So, when man came to understand the regular movements of the stars and could forecast them accurately, he tried to bring all life under their sway; and the science of astronomy became the superstition of astrology. A similar tendency accompanied the progress of geography. A knowledge of the climates and characters of the different regions of the world gave birth to the idea of explaining the histories of those regions as effects of geographical causes. Hot climates produce lazy men, cold ones stupid men, temperate ones active men. Plains have one effect, mountains another, and so on. Blanket generalisations of this kind were applied to the neglect of the facts. Geographical determinism threatened the study of history, and it became necessary to protest that a great nation was a human masterpiece, not a natural product; that it was the fruit of innumerable human decisions, of well-weighed actions, of a steeled will, of creative intelligence, in short, of a struggle with the environment and a conquest of it, not a passive submission to imagined geographical causes.

So it was that in Darwin's day both Darwin himself and other thinkers, impressed by the biological theory of natural selection and the doctrine of the survival of the fittest, thought it could be extended to cover also the sphere of human history. Darwin's imagination had been filled with the spectacle of raised continents and submerged ocean beds. He had seen the fossil remains of innumerable species of living things destroyed by natural causes. He had explained the creation of vast natural features like the coral reefs of the Pacific by the mindless activity of countless generations of myriads of creatures of the lowest grade of life. He had devoted his life to disentangling the reign of law from the confusion of the phenomena and had at length arrived at the theory of natural selection. Naturally enough, when he turned his mind also to the human scene, he sought to apply the same principle.

It was not, of course, that he was unaware of the difference between the world of nature and the world of man. But he did not adequately measure it. Let us repeat some words of his from the *Descent of Man:* "Man in the rudest state in which he now exists is the most dominant animal that has ever appeared on earth. . . . He manifestly owes his immense superiority to his intellectual faculties, his social habits, which lead him to aid and defend his fellows, and to his corporeal structure." Here he distinguishes intellect, morality, and corporeal structure. But when he passes on to explain their transmission and development from generation to generation, all alike are regarded as part of man's biological inheritance. "Tribes, including many members who were always ready to give aid to each other and to sacrifice themselves for the common good, would be victorious over most other tribes; *and this would be natural selection.* At all times throughout the world tribes have supplanted other tribes; and as morality is one element in their success, the standard of morality and the number of well-endowed men will thus everywhere tend to rise and increase."

This is sad stuff. Patriots do not necessarily beget patriots. There is no gene for this virtue. Moral progress is not achieved in this mindless way. Nor did Darwin really think so. But he had no philosophy which could provide him with any other reasonable account of the true nature of the mental world. After all, he had been satisfied, and more than satisfied, with his bold speculation that "thought is a secretion of the brain." The brain is certainly biologically inherited. The propositions that "It is sweet and noble to die for one's country" and "Greater love hath no man than this that a man lay down his life for his friend" are certainly thoughts. Why then should not a brain that had secreted these thoughts in one generation secrete them also in the next? But this, of course, is nonsense. Such thoughts are part of the cultural history of mankind, which is superimposed on the biological and not to be identified with it. Darwin had failed to make a clear distinction between the brain and the mind, and the failure had a disastrous effect on his mental life and happiness.

Darwin and the Poets

When a great advance in scientific knowledge is made it has the effect of casting the culture of preceding ages temporarily into the shade. "They did not even know that the earth goes round the sun," we say of them; "why bother about them anymore?" But this is a mistake. There are great areas of culture that are not affected by revolutions in our theories about the world of nature. The thought and art of ancient Athens, or medieval Paris, or renaissance Florence, though we may see it in a different perspective, has not been diminished or dimmed by the Copernican revolution.

A blindness of this sort seems to have afflicted Darwin after his discovery of the law of natural selection. "The old argument of design in nature, as given by Paley, which formerly seemed to me so conclusive, fails, now that the

law of natural selection has been discovered." So he wrote
in his *Autobiography;* and so far, so good. Nobody would
want him to deny evolution, as Galileo was made to deny
that the earth moves. There was nothing to be done with
Paley but to dismiss him. The trouble began when Darwin, absorbed in elaborating his doctrine of natural selection, lost interest also in the wider culture which had once
delighted him.

Milton, for example, had been a passion with the youthful Darwin. *Paradise Lost* was one of the few books he
took with him on his voyage to South America. When he
went off on his long and lonely expeditions into Argentina
or Patagonia it was his Milton he had in his pack. Now
Milton, of course, was a believer in a once-for-all creation
of the animal kingdom; and he wrote about it in a style
more elevated than Paley but in essence no less absurd.

> God said
> Let the Earth bring forth fowl living in their kind,
> Cattle and creeping things, and Beast of the Earth,
> Each in their kind. The Earth obeyed. . . .
> The grassy clods now calved; now half appeared
> The tawny lion, pawing to get free
> His hinder parts . . .
> the swift stag from underground
> Bore up his branching head. . . .
> *Paradise Lost,* vii, 450 ff.

It is ironical to think that *Paradise Lost* was Darwin's
campfire reading on that wonderful expedition to Punta
Alta when he discovered the fossilised remains of nine
great extinct quadrupeds which led to a decisive step forward in his theory of natural selection. The goddess of

luck had great sport with Darwin on that occasion. If her intention had been to cure him of his taste for poetry, no better means could have been devised. There was Nature, on the one hand, offering her choicest exhibits to the budding evolutionist, and poetry, on the other, with nothing better to show than a grandiose and rhetorical elaboration of the kind of error he had begun to detest in Paley. Darwin does not tell us that Punta Alta destroyed his taste for *Paradise Lost*. But this, or something like it, is what the facts reveal.

In his *Autobiography* he notes that from about 1842 on he began to lose interest in scenery, poetry, and even in his friends. "This curious and lamentable loss," he writes, "is all the odder, as books on history, biographies, and travels (independently of any scientific facts which they may contain), and essays on all sorts of subjects, interest me as much as ever they did. My mind seems to have become a machine for grinding general laws out of large collections of facts, but why this should have caused the atrophy of that part of the brain alone, on which the higher tastes depend, I cannot conceive." Again, about scenery, with special reference to the Brazilian forest, he writes: "It is not possible to give an adequate idea of the higher feelings of wonder, admiration, and devotion which fill and elevate the mind." But he adds: "I cannot see that such inward convictions and feelings are of any weight as evidence of what really exists."

Perhaps in the last phrase, "evidence of what really exists," we can find a clue to what was happening to Darwin. He talks much of the "higher feelings," but he cannot take them for what they are. He wants them to be "evidence" of something else; and the something else, of which they are not evidence, is "what really exists." Obviously this is a limitation of the sphere of "real existence" to what can be proved by extracting generalisations from large collections of facts. A law like natural selection really exists. But delight in natural beauty or delight in poetry do not prove anything. True, they are in some way "higher" than the findings of natural science. But since they are not evidence of the truth of some natural law they are tainted with unreality. Such is the conclusion in which Darwin finds himself, much to his regret, compelled to acquiesce.

"If I had my life to live over again I would make a rule to read some poetry or listen to some music at least once a week; for perhaps the parts of my brain now atrophied could thus have been kept active through use. The loss of these tastes is a loss of happiness, and may possibly be injurious to the intellect, and more probably to the moral character, by enfeebling the emotional part of our nature." So he ruminates, and his ruminations only show how far he had gone off the track. To those who care for poetry or music or painting or philosophy or religion, these things are not "higher tastes" which exist on the periphery of reality. They are not solaces and adornments which, for half an hour "at least once a week," refresh the scientist who alone is concerned with "what really exists." Music, poetry, painting, philosophy, religion are concerned with what really exists and give knowledge of what really exists. And the scientist needs them because they rescue him from the abstractions—the atoms, the genes, or whatever they may be—which make up the content of his particular study. They refresh and restore him by bringing him into contact with aspects of reality with which his specialism cannot deal.

Poetry, in fact, is much less abstract than biology. Darwin's loss of interest in Shakespeare illustrates this fact.

When he was young he had a taste for Shakespeare, especially the Histories. Now these plays are without parallel
in literary history for the solidity with which they evoke
the image of a people's life, high and low, in town and
country, in court and camp, in peace and war, during two
hundred vital years of history. Those who retain interest
in life retain interest in the plays, for they enlarge the understanding, quicken the perceptions, expand the sympa-
thies in a unique degree. They are unsurpassed in their
grasp of concrete social reality.

It was Darwin who was becoming more and more abstract in his approach to society, being occupied only with
the one concern of proving that everything social could be
interpreted as a fresh example of the law of natural selection. Pleasures, he tells us in his *Autobiography*, have a survival value and so are preserved biologically. "Hence it has
come to pass," we read, "that most or all sentient beings
have been developed in such a manner through natural selection, that pleasurable sensations serve as their natural
guide. We see this in the pleasure from exertion, even occasionally from great exertion of the body or mind—in
the pleasure of our daily meals, and especially in the pleasure derived from sociability and from loving our fami-
lies." (page 89.) Biologically this is nonsense. There is no
gene for domestic felicity. Do not look to natural selection
to secure you that. And, in addition, what a degree of ab-
straction! When the word pleasure is stretched to cover
physical sensations, the satisfaction of great mental exertion, and the happiness of being with family and friends,
it has become so tenuous as to be almost without meaning.

It would be pleasant if we could leave the matter here.
Unhappily this attempt of Darwin's (in which he was not
alone) to bring the phenomena of society under the bio-

logical law of the survival of the fittest offered a scientific justification for some of the vilest acts in history. Social Darwinism, as it has been called, is an ugly chapter in our recent past. If the survival of the fittest is nature's law and if nature's law is to rule also in society, then there is a case for sending the weakest to the wall, or, to put it bluntly, for genocide. The argument was not lost on the Nazis.

But we may drop the subject here. Nothing could be less in keeping with Darwin's personal character than racial antipathy, enslavement of weaker peoples, or extermination of those judged by butchers to be less fit. *The Voyage of the* Beagle, which when all is said and done remains his best book, is full of generous and just comment on all sorts and conditions of men. All that we need to uncover and reject is Darwin's inability to distinguish a human from a biological fact.

Darwin and Christianity

Interest in Darwin's rejection of Christianity was revived a few years ago by the publication of the complete text of the *Autobiography* (© Nora Barlow, 1958). He describes the change of opinion that began to overtake him in 1836 on his return from the voyage of the *Beagle:*

> I had gradually come to see that the Old Testament from its manifestly false history of the world, with the Tower of Babel, the rainbow as a sign, etc., etc., and from its attributing to God the feelings of a revengeful tyrant, was no more to be trusted than the sacred books of the Hindoos, or the beliefs of any barbarian. . . .
>
> By further reflecting that the clearest evidence would be required to make any sane man believe in the miracles by which Christianity is supported . . . I gradually

came to disbelieve in Christianity as a divine revela-
tion. . . . Disbelief crept over me at a very slow rate,
but was at last complete. The rate was so slow that I felt
no distress, and have never since doubted even for a
single second that my conclusion was correct. (page
85–87.)

We might connect his loss of faith with the other symp-
toms of mental atrophy of which he complains. But this
would be quite unfair. It is more to the point to recall the
limitations of the orthodox tradition in which he was
brought up. We have to remember that these limitations
were such that even as a student of divinity at Cambridge
he could still profess belief in "the strict and literal truth
of every word of the Bible." Considering the blinkers with
which he had been fitted we must give him credit for seeing that the Tower of Babel and Noah's Ark are not the
same kind of thing as the histories his contemporary,
Macaulay, was writing. But all the same the medieval
guildsmen who handled these old oriental tales with imaginative freedom in their mystery plays had more sense
than this. They did not bother about their literal truth,
but felt instinctively their imaginative truth as a comment
on human life.

Darwin's complaint about the revengeful feelings at-
tributed to God does credit to his heart but springs from
the same cultural inadequacy. Nowadays the evolution of
the concept of God in the Old Testament, the transfor-
mation of a tribal god of a desert people into the father of
all mankind proclaiming "I will have mercy and not sacrifice," is recognized as a unique achievement of the Hebrew genius for religion. But, as we have seen, Darwin,

whose understanding of cultural history was very limited, regarded what he called "the ennobling belief in an omnipotent God" as a product of biological evolution.

In his criticism of the New Testament we sympathise with his revolt from a view of Christianity which relies on miracles as evidence of its truth. All these breaches of the natural law serve only to obscure the great miracle, the Gospel itself. Darwin threw the baby out with the bathwater, as did many of the best of his contemporaries. But he left us with a new riddle—how to explain the Sermon on the Mount by natural selection.

But his achievement was not merely negative. The fact is that Darwin in the nineteenth century was performing a service like that of Galileo in the seventeenth. He was forcing the orthodox to revise their attitude to their sacred books. The effect of the two men's work has been permanent. They brought to an end the practice of using the Bible as an authority on physical and biological science—a thoroughly wholesome development. But it has had an unfortunate side effect. For a certain type of mind, increasingly common in our age, whose whole conception of truth is limited to the natural sciences, the Bible has become a closed book. Its kind of truth has become invisible.

Science claims, and rightly claims, to be ethically neutral. It is easy, then, to understand that the great modern advances in science, while they increase our knowledge, do not deepen our wisdom. Copernicus revived an old Greek theory that the earth goes round the sun. Newton discovered the law of gravity. Harvey discovered the circulation of the blood. Darwin discovered the principle of natural selection. Dalton revived the atomic theory of Democritus. The physicists split the atom, which was by definition unsplittable. And so on. These advances have so altered our conception of the external world that we are inclined to think, and are often told, that the world was plunged in darkness before they were made. But we forget that great civilizations, in many ways superior to our own, flourished before the natural sciences in their present phase were born.

Men did not need to prove that the earth goes round the sun in order to create Greek literature or Roman law, or—what is more fundamental—to create the Greek city-state and the Roman empire, which were the social bases of the literature and the law. Shakespeare knew nothing of Harvey's discovery about circulation or Gilbert's about the magnet. But that did not prevent him from becoming England's, if not the world's, greatest dramatic poet, i.e., from understanding social reality more deeply than any other man.

And so with the Bible. The old Hebrew writers have no vocabulary in which to deal with the laws of nature. The very concept of nature, which we have borrowed from the Greeks, was unfamiliar to them. They were concerned with something different, but certainly not less real. We may call it wisdom rather than science. In their view the ultimate reality was moral, a moral being who could only be understood by those who were trying to do his will. This too is a point of view, but one Darwin missed.

But in abandoning Christianity Darwin did not become an atheist; he reverted to the theism of his father and grandfather. He chose always to lard his writings with references to the First Cause or the Creator. The difficulty is to discover what he meant by these terms. He seems to have recognised two kinds of reasons for believing in

God, the emotional and the rational. He discusses the first in relation to the tremendous impression made upon him by the Brazilian forest. But, as we have seen, he dismissed this with the remark: "I cannot see that such inward convictions and feelings are of any weight as evidence of what really exists." He then proceeds to deal with the possible rational grounds for belief:

> Another source of conviction in the existence of God, connected with the reason and not with the feelings, impresses me as having much more weight. This follows from the extreme difficulty or rather impossibility of conceiving this immense and wonderful universe, including man with his capacity for looking far backwards and far into futurity, as the result of blind chance or necessity. When thus reflecting I feel compelled to look to a First Cause having an intelligent mind in some degree analogous to that of man; and I deserve to be called a Theist. (*Autobiography,* page 92. © Nora Barlow 1958.)

This conclusion, he adds, was strong in his mind when he wrote the *Origin of Species,* but gradually became weaker. His correspondence gives further evidence of the fluctuations and weakening of his faith:

> The impossibility of conceiving that this grand and wondrous universe with our conscious selves arose through chance, seems to me the chief argument for the existence of God; but whether this argument is of real value, I have never been able to decide. . . . The safest conclusion seems to be that the whole subject is beyond the scope of man's intellect.

And again:

> In my most extreme fluctuations I have never been
> an atheist in the sense of denying the existence of a
> God. I think that generally (and more and more as I
> grow older), but not always, an agnostic would be the
> more correct description of my state of mind.

In another letter, with charming candour, he calls his theology "a simple muddle." (Gavin de Beer, page 268.)

Religion, however, continued to occupy his mind and
probably the best-thought-out and clearest statement of
his views is to be found in the conclusion of *The Variation
of Plants and Animals Under Domestication:*

> In accordance with the views maintained by me in
> this work and elsewhere, not only the various domestic
> races, but the most distinct genera and orders within
> the same great class—for instance, mammals, birds,
> reptiles, and fishes—are all the descendants of one
> common progenitor, and we must admit that the whole
> vast amount of difference between these forms has pri-
> marily arisen from simple variability. To consider the
> subject from this point of view is enough to strike one
> dumb with amazement. But our astonishment ought to
> be lessened when we reflect that beings almost infinite
> in number, during an almost infinite lapse of time, have
> often had their whole organisation rendered in some
> degree plastic, and that each slight modification of
> structure which was in any way beneficial under exces-
> sively complex conditions of life has been preserved,
> whilst each which was in any way injurious has been
> rigorously destroyed. And the long continued accumu-
> lation of beneficial variations will infallibly have led to

structures as diversified, as beautifully adapted for various purposes and as excellently co-ordinated, as we see in the animals and plants around us. Hence I have spoken of selection as the paramount power, whether applied by man to the formation of domestic breeds, or by nature to the production of species. . . .

And here we are led to face a great difficulty, in alluding to which I am aware that I am travelling beyond my proper province. An omniscient Creator must have foreseen every consequence which results from the laws imposed by Him. But can it be reasonably maintained that the Creator specially ordained for the sake of the breeder each of the innumerable variations in our domestic animals and plants—many of these variations being of no service to man, and not beneficial, far more often injurious, to the creatures themselves? Did He ordain that the crop and tail-feathers of the pigeon should vary in order that the fancier might make his grotesque pouter and fantail breeds? Did He cause the frame and mental qualities of the dog to vary in order that a breed might be formed of indomitable ferocity, with jaws fitted to pin down the bull for man's brutal sport? But if we give up the principle in one case . . . no shadow of reason can be assigned for the belief that variations, alike in nature and the result of the same general laws, which have been the groundwork through natural selection of the formation of the most perfectly adapted animals in the world, man included, were intentionally and specially guided.

We content ourselves with one comment on this carefully considered passage. Here, as everywhere else, Darwin proceeds on the analogy between the work of the plant-and-animal breeder and the work of nature. The breeder selects the variations he wants, breeds only from them, and destroys the rest. So natural selection, personified as "the paramount power," "preserves" beneficial variations and "rigorously destroys" those even slightly injurious, and so on. But "selection" is only a metaphor transferred from the conscious, purposive activity of the human breeder to the unconscious, purposeless activity of nature. It would be less misleading to describe nature's process as a sort of sieve which automatically sorts out the different constituents in a mixture. It does not justify Darwin's terminology. It does not justify the title of his chief work: *The Origin of Species by Means of Natural Selection or the Preservation of Favoured Races in the Struggle for Life*. There are no favoured races and no selection. These phrases belong to the vocabulary of conscious design, which is precisely what Darwin wishes to exclude.

Darwin's Place in
the History of Thought

Evolution in the broadest sense of the word has three phases: inorganic, organic, and human. In the inorganic phase the only laws that operate are physico-chemical. The moon, when we get there, should offer an example of this phase. A lifeless orb.

Then, perhaps about two thousand million years ago, began on earth the second phase, the organic. This is the domain of biological law. So far as we yet certainly know, this phase has occurred nowhere else. During this phase there was plant and animal life without man. What the earth was then like can be inferred from the fossil record and experienced in a few unpeopled regions. Its image may be to some degree perpetuated for future generations by nature reserves.

When did the third phase, the human, begin? If we look for traces of man as a tool-maker, possibly about two million years ago; if we look for *homo sapiens,* perhaps thirty thousand years ago; if we choose the time when man became truly the dominant creature, then perhaps no more than ten thousand years ago. It is astonishing how quickly life, once it began, spread in its myriad forms all over the globe in sea, land, and air. It is equally astonishing how man, since he emerged as the dominant species, has done the same.

The third, the human, phase is the domain of conscious, purposive life, the psycho-social phase. It is in this third phase that Darwin walks with an uncertain step. Indeed he never clearly recognised or acknowledged its separate existence. It was for him a mere extension of the biological phase. His subject is the origin of species, not cultural history.

In the first two evolutionary stages Darwin was a pioneer. His quick insight into nature's ways enabled him to anticipate modern views on the transition from the inorganic to the organic stage and on the dramatic change in the whole economy of nature involved in the advent of life. Sir Gavin de Beer quotes from a letter of 1871:

It is often said that all the conditions for the first production of a living organism are now present. But if (and Oh! what a big if!) we could conceive in some little pond, with all sorts of ammonia and phosphoric salts, light, heat, electricity, etc., present, that a protein compound was chemically formed ready to undergo still more complex changes, at the present day such matter would be instantly devoured or absorbed, which would not have been the case before living creatures were formed.

But no such flash of insight illuminated for Darwin the magnitude of the transition from the animal to the human, from life to self-conscious life. He was better equipped to understand nature than human nature; and, after all, the urgent necessity in his day was to demonstrate the origin of species, including man, by natural selection. This, in spite of illness and opposition, Darwin did. It was for a later generation to rediscover, and redefine, the uniqueness of man.

In the early days of British science, before the foundation of the Royal Society, Francis Bacon had urged his generation to give up spinning theories of the universe out of books and to go straight to nature herself. This he called "the commerce of the mind with things". Nobody ever followed this advice with more devotion than Darwin. But Bacon had added a caution which Darwin, if he had ever heard of it, did not heed. It was a weakness of the human mind, Bacon warned, when it had discovered a truth by concentration on one field, to think it capable of universal application. Darwin exemplifies this tendency. Having made an immense advance in understanding the mode of evolution in the instinctive realm of animal life, he then without any misgiving applied his findings to the rational life of man.

But these two faculties, instinct and reason, differ so much that what is true of the one cannot be transferred without more ado to the other. To ignore the difference, as Darwin did, was to blind himself to the whole history of mankind. Man is what he is on account of the institutions he has created. In the animal world instinctive patterns of behaviour are handed down from generation to generation by the mechanism of biological inheritance. This is true even of those patterns of behaviour which most resemble human communication, like the warning cries of animals and the dancing of bees to indicate the location of sources of honey. But in human society the transmission of the cultural inheritance is not biological but educational. The education consists in a conscious initiation into all forms of social life, and the essential medium of the initiation is language.

Darwin, of course, knew this, but the knowledge played no part in the formation of his theory of evolution as applied to man. He speculates, for instance, on the changes necessary to turn a naked Fuegian into a civilized man, and it is plain that he understands them in a biological rather than a cultural sense. This is a new kind of misunderstanding such as often arises from the progress of knowledge. Two hundred and fifty years earlier Francis Bacon was free from it. Writing at the time of the first contact between the English and the American Indians he says: "Consider the abyss which separates the life of men in some highly civilized region of Europe from that of some savage, barbarous tract of New India. So great is it that one might appear a god to the other. . . . And this is the effect not of soil, not of climate, not of physique, but of the arts." In other words, while Bacon would educate the Fuegians, Darwin would wait for them to evolve. Natural selection, not education, would have to do the job. We have already insisted that Darwin personally was a notably humane and friendly man. But his theory can still be invoked to lend a scientific colour to those oppressive ruling minorities in various parts of the world who deny rights to their less educated subjects on the grounds that they are not yet sufficiently evolved.

In comparison with modern conceptions of evolution

Darwin's inclusion of the human in the biological stage is the greatest weakness of his work. Naturally modern sociologists are not slow to dissociate themselves from this outmoded view. The essential point is put succinctly by Ginsberg: "Mental development in man is a social process." We can set aside Darwin's unhappy biological image. Thought is not a secretion of the brain, but the creation of man in society. Consequently any analogy between biological and sociological processes is misleading. "No analogy between the evolution of species and the evolution of societies is valid," warns Gordon Childe.

Neither Ginsberg nor Childe doubts the evolution of the human *brain* from the animal. The point is the recognition of the new organ, the mind. The mind cannot exist without the brain. But the brain marks the end product of biological evolution and the mind the beginning of social, human, evolution. So Durkheim asserts: "Sociology places itself from the start in the ideal sphere; it does not arrive there by slow degrees, as the end of its researches; it sets out from them. The ideal is its proper domain." Finally, to quote a contemporary evolutionist who successfully bridges the gap between the biological and the social, Sir Julian Huxley writes: "The brain alone is not responsible for mind, even though it is a necessary organ for its manifestation. Indeed an isolated brain is a piece of biological nonsense, as meaningless as an isolated human individual." The implications of these pronouncements for modern evolutionary theory are immense. For while biological selection pushes evolution on mechanically from behind, psycho-social selection, which involves conscious aim and purpose, draws man from in front.

The clear distinction between the biological and the psycho-social worlds is the greatest advance in the general

theory of evolution since Darwin. But in the purely biological sphere also Darwin's work has been superseded by
the experiments of Mendel and the rise of the new science
of genetics. One might have expected Darwin himself to
be the founder of modern genetics. But the whole history
of science teaches us not to be surprised when a man who
has made a great advance proves incapable of taking the
next step. Here the comparison between Mendel's success
and Darwin's failure is instructive.

Mendel, we are rightly reminded, had the advantage of
reading the *Origin of Species*. It was this that interested
him in the problem of inheritance. He owed a debt to
Darwin. But this only serves to emphasise the point,
which is that both men had the same evidence before
them, and Mendel saw how to make use of it while Darwin did not.

We have already briefly described the series of experiments made by Mendel in the 1860s. The secret of
Mendel's success lay in his capacity to tackle the problem
of inheritance analytically. He chose easily recognizable
and clearly definable characters; he made each character
the subject of a special investigation; he applied the precision of number to his results; his seven parallel series of
experiments on separate characters supported and confirmed one another. In fact, the only mistake Mendel
made was not to be already famous and consequently to
have published his findings in an obscure journal which
remained unread for over thirty years!

At about the same time Darwin was engaged in a similar investigation but was unable to bring it to a satisfactory
conclusion. The evidence at his disposal is set forth in the
two large volumes of his *Variation of Animals and Plants
under Domestication*. It was not the evidence that was lack-

ing but the flair to find an experimental method of interpreting it. The idea at the back of Darwin's mind was that in sexual generation the qualities of the parents are blended in the offspring. The inability to shake off this popular idea of "blending inheritance" stood in his way. As he piled up his evidences he came across a number of exceptions to what he took to be the rule. These he grouped under the heading: *On Certain Characters not Blending.* But their significance was lost on him. He did not quite know what to do with them, so he had them printed in small type to satisfy his scientific conscience. There they stand in his book to this day to witness to the fact that he rejected the stone that was to become the head of the corner. Here are some examples from his list:

> Some characters refuse to blend, and are transmitted in an unmodified state either from both parents or from one. When grey and white mice are paired, the young are piebald, or pure white or grey, but not of an intermediate tint; so it is when white and common turtledoves are paired. In breeding Game Fowls, a great authority, Mr. J. Douglas, remarks, "I may here state a strange fact: if you cross a black with a white game, you get birds of both breeds of the clearest colour." Sir R. Heron crossed during many years white, black, brown, and fawn-coloured Angora rabbits, and never once got these colours mingled in the same animal, but often all four colours in the same litter.

So much for animals; plants yielded the same results:

> Gaertner crossed many white and yellow-flowered species and varieties of Verbascum; and these colours

were never blended, but the offspring bore either pure white or pure yellow blossoms; the former in the larger proportion. Dr. Herbert raised many seedlings, as he informed me, from Swedish turnips crossed by two other varieties, and these never produced flowers of an intermediate tint, but always like one of their parents. I fertilised the purple sweet-pea (*Lathyrus odoratus*), which has a dark reddish-purple standard-petal and violet-coloured wings and keel, with pollen of the painted-lady sweet-pea, which has a pale cherry-coloured standard, and almost white wings and keel; and from the same pod I twice raised plants perfectly resembling both sorts; the greater number resembling the father. So perfect was the resemblance, that I should have thought there had been some mistake, if the plants which were at first identical with the paternal variety, namely, the painted-lady, had not later in the season produced flowers blotched and streaked with dark purple. I raised grandchildren and great-grand-children from these crossed plants, and they continued to resemble the painted-lady, but during later genera-tions became rather more blotched with purple, yet none reverted completely to the original mother-plant, the purple sweet-pea.

Here, partly from Darwin's sources and informants, and partly from his own experiments, we get closer and closer to Mendel's classic work, but without Mendel's in-sight. The analysis lacks precision, the experiments lack direction. The distinction between the external appear-ance and the genetic makeup, that is to say, between the phenotype and the genotype, to use the current terminol-ogy, is not clearly drawn. Nor is the precision of number introduced. It was left to Mendel to apply the statistical method to a biological investigation. Darwin contents himself with phrases like "the greater number," "the larger proportion." So the crucial significance of the results is missed. They are treated as curiosities or abnormalities which cannot be fitted in to the old, popular, and established view of "blending inheritance." It is curious to reflect that when Darwin brought out his two large volumes on *Variation* in London, Mendel's epoch-making paper, which renders Darwin's conclusions out-of-date, had already been published in the proceedings of the local naturalists' society in Brunn.

But Darwin's failure to anticipate, or at least tie with, Mendel was hardly an accident. It originated from certain faults of method plainly apparent in his most famous book. The *Origin of Species,* though fully deserving of its great place in the history of science, lacks some of the essential qualities of a scientific masterpiece. However we seek to excuse it, the absence of any attempt to place his theory in its historical setting is a grave defect; and the historical sketch which he did prefix to the edition of 1861 does less than nothing to repair the omission. For a scientist of the highest rank it is essential that he be capable of a just estimate of his own place in an historical succession. He should be able to make due acknowledgment to his predecessors, and to define his own contribution by setting it in a true relation to theirs. Here Darwin failed conspicuously. He makes it virtually impossible for the reader to know for what elements in the theory he claims originality. So much so, that one is haunted by the suspicion that Darwin had never made it clear to himself. For the student who wishes to understand the history of evolutionary theory there could be no more misleading work than the *Origin of Species.*

Before Charles Darwin was born the concept of descent with modification was well established. Buffon, Erasmus Darwin, and Lamarck had all recognised it; and they had tried, with varying degrees of success and with shifts of emphasis from one possible solution to another, to suggest how the emergence of new species might result from the response of living things to the changes in their environment. Charles Darwin's great originality lay in directing attention to the spontaneous variations, the "sports," which arise from time to time, and in revealing the action of natural selection in preserving favourable variations. But a full and careful analysis of the contribution of the pioneers would not only have done justice to them but clarified his own position.

No reader, however, could guess from the opening page of the *Origin* that descent with modification had a long history before Darwin took up his pen. What he there tells us is that on the voyage of the *Beagle* he "was much struck with certain facts in the distribution of organic beings inhabiting South America." "These facts," he adds, "seemed to throw some light on the origin of species— that mystery of mysteries." On his return home, he goes on, "it occurred to me that something might perhaps be made out on this question by patiently accumulating and reflecting on all sorts of facts. . . . After five years' work I allowed myself to speculate on the subject." That is all. Not a word about Buffon, grandfather Erasmus, or Lamarck; not a word about the surprising Mr. Patrick Matthew, who, in his book *On Naval Timber and Arboriculture*, published in 1831, speaks in a long and forceful passage of "the immense waste of primary and youthful life" in nature, which brings it about that "those only come to maturity" who have survived "the strict ordeal by which nature tests their adaptation to her standard of perfection and fitness to continue their kind by reproduction". The subject was in the air, and Darwin does not say so.

Inevitably, then, when we come to the end of the book and Darwin sums up the main laws of evolution, we think he is telling us what he had worked out for himself by pondering on the observations he had made during the voyage of the *Beagle*. He lists the main laws of evolution. They are: "(1) Growth with Reproduction; (2) Inheritance which is almost implied by reproduction; (3) Variability from the indirect and direct action of the conditions of life, and (4) from Use and Disuse; (5) a Ratio of Increase so high as to lead to a Struggle for Life, and as a consequence (6) to Natural Selection, entailing Divergence of Character and the Extinction of less-improved forms." We think this is all the result of Charles Darwin's own cogitations because he has not plainly told us that (1), (2), (3), and (4) were anticipated by one or more of his predecessors. Nor has he told us that (5)—Malthus's idea about the struggle for survival—had suggested (6)—the idea of natural selection—also to his friend Wallace. All this could and should have been set forth clearly; and it would have added to rather than detracted from our appreciation of his own achievement. For to Darwin rightly belongs the credit of having supported the theory of evolution with such a mass of evidence as to ensure its acceptance.

But while it is proper to give Darwin his due, it is equally necessary not to lose sight of his limitations. These explain not only, what we have already discussed, his blindness to the distinction between the biological and the social worlds. They explain also, I think, how he came to miss the significance of the facts which led Mendel to the discovery of particulate inheritance and made him the true founder of genetics. What I mean is this. Just as a true picture of the history of evolutionary theory was beyond Darwin's powers, because, as the whole course of the argument in the *Origin* reveals, he never managed to keep clearly in mind the precise contribution of Buffon, Erasmus Darwin, and Lamarck to the formation of that theory, so he lacked the capacity to make the nice distinctions without which Mendel could not have planned his classical experiments.

In Darwin's day biological inheritance was popularly understood to mean the handing on of similar characters from parents to offspring. But "character" is a very vague word. It could mean anything from a long nose to heroic self-sacrifice. All sorts of qualities and patterns of behaviour, physical, mental, and moral, were included in the term. How this looseness might lead to looseness of thought may be understood from a passage in the *Variations of Animals and Plants under Domestication*:

How can we explain the inherited effects of the use and disuse of particular organs? The domesticated duck flies less and walks more than the wild duck, and its limb-bones have become diminished and increased in a corresponding manner in comparison with those of the wild duck. A horse is trained to certain paces, and the colt inherits similar consensual movements. The domesticated rabbit becomes tame from close confinement; the dog, intelligent from associating with man; the retriever is taught to fetch and carry; these mental endowments and bodily powers are all inherited. Nothing in the whole circuit of physiology is

more wonderful. How can the use or disuse of a particular limb or of the brain affect a small aggregate of
reproductive cells, seated in a distant part of the body,
in such a manner that the being developed from these
cells inherits the *characters* of either one or both parents. . . . We may on the whole conclude that inheritance is the rule, and non-inheritance the anomaly.
(pages 367–68.)

The first thing we may notice about this paragraph is
that it is pure Lamarckism. The characters discussed are
all acquired characters and it is taken for granted that they
are inherited. "These mental endowments and bodily
powers are *all* inherited." "Inheritance is the rule." And
next we may observe the wide connotation of the word
"character." The shrinkage of the wing bones and the
growth of the leg bones in the domesticated duck through
disuse or use; the paces acquired by the horse under training; the tameness of the confined rabbit; the intelligence
the dog acquired by association with man; the retriever's
willingness and skill at fetching and carrying—all these
are "characters." Darwin is here defining the problem
which he is about to answer by his theory of Pangenesis.
The task he sets himself is to frame some hypothesis to
explain how these immensely complex mental endowments and bodily powers are transmitted through the reproductive cells.

The terms of the problem are too loosely defined to be
susceptible of experimental test. Contrast the procedure
of the successful Mendel. He selected a few simple,
clearly defined, physical traits for experimental investigation. Peas are round or wrinkled, yellow or green. But who
can measure accurately the tameness of a rabbit or the in-

telligence of a dog? On a problem so loosely formulated experimental investigation will yield no clear answer. Speculation is all that is possible, and speculation is all we get in Darwin's theory of Pangenesis. The theory is elaborate, but roughly his idea is that every cell in the body casts off innumerable little buds, which he proposed to call "gemmules." These gemmules embody the experience gained by every cell of the body during the course of its life. These experience-laden gemmules, coming from every cell in the body, somehow make their way into the reproductive cells, where Darwin admits there may well be a problem of overcrowding, and there they remain ready to play their specific role in the production of the new individual. In this way acquired characters are inherited. In this way we may, albeit dimly, conceive how the paces of the trained horse and the retriever's gift of fetching and carrying may be transmitted to the colt and the puppy.

"Even an imperfect answer to this question of the inheritance of acquired characters," wrote Darwin hopefully, "would be satisfactory." Satisfactory, in the event, proved to be an ill-chosen word. The theory of Pangenesis was badly received in its own day. It was even unkindly pointed out that the theory, in a sketchy form, had been anticipated by one of the Hippocratic writers of ancient Greece more than two thousand years earlier. Nor is this fact without significance. Ancient Greek science was in the main speculative for want of an established practice of experiment, and in the matter of genetics Darwin was in no better case. He had been unable to discover a method of experimental investigation in this field.

There is no reason, then, to be surprised that Pangenesis is the reverse of modern genetic theory. According to Pangenesis it is the whole body which issues instructions to the reproductive cells. In the modern theory it is the fertilised ovum which issues instruction for the making of the whole body.

It may seem unkind to bring our account of Darwin's work to a close by dwelling on the unsuccessful theory of Pangenesis. But our purpose in this chapter is not to list Darwin's successes, which has been done elsewhere in the book, but to assess his place in the history of thought. His *Origin of Species* proved the turning point in a long struggle for a new conception of natural history. It was decisive in replacing the static conception of a universe created-once-for-all by the conception of a universe in a state of evolution. And it proved decisive by reason of solid merits, namely the control of a vast body of material largely collected by himself and marshalled with great skill. By sweeping away the type of argument used by Paley it forced us to attend to an aspect of reality previously ignored and transformed the whole mental landscape.

But the struggle had been a long struggle, and Darwin gives us no understanding of its history and very little insight into the nature of the religious and philosophical problems involved. His lack of historical sense, deriving from a general lack of culture, is a grave defect. His style, though not lacking in freshness and naturalness, is that of an undisciplined thinker not practised in drawing fine distinctions. His enthusiasm for nature is infectious, but when he strays beyond the bounds of natural philosophy, his opinions are feeble and fluctuating. More often than not he resorts to hollow-sounding platitudes which are a poor substitute for firm convictions. One exempts from these charges the ever-fresh and ever-interesting *Voyage of the* Beagle, which the Duke of Argyll rightly calls "the most delightful of all his books." But in no sense is he a classical writer. Patient, incredibly industrious, a great observer of nature, yes. But not a profoundly original thinker, not a great mind.

Nor should we leave the matter here. Wherein lies the difference between the *Voyage of the* Beagle and the later books? Is the Duke of Argyll perhaps right that the difference is that in the *Voyage* we have "Darwin before he became a Darwinian"? For Darwin had not then raised for himself all the problems that arise from his special brand of evolution. If, as he maintained when he was not simply repeating Lamarck, the whole process of nature is nothing but the result of the mindless action of natural selection on the random variations that arise between one generation and the next, how then account for the life of mind and purpose for whose manifestations he showed such generous regard in his first book? These were the questions the mature Darwin was half-afraid to look at and wholly incapable of answering. If he had been able to raise them sharply instead of trying to pretend they do not exist, his books would have had more of the breath of life and more claim to survival. Instead he tended to regard the life of the mind as somehow tainted with unreality. But this is an unwarranted conclusion. The process of evolution, which has produced life out of matter, and thought out of life, is not a diminution but a growth of reality; and the Darwinian account of biological evolution, so far as it is true, is not a complete account of reality, but a partial account of one aspect of the biological process. The whole realm of truth is not covered by the theory of natural selection.

Bibliography

A. Works by Charles Darwin:
   1. *Journal of Researches into the Natural History and Geology of the Countries visited during the Voyage of the* Beagle *round the World*. First publ., 1839. Ed. used: T. Nelson and Sons, London, 1890.
   2. The *Origin of Species by Natural Selection or The Preservation of Favoured Races in the Struggle for Life*. First ed., 1859. Ed. used: Reprint of Sixth Edition (1872), John Murray, London, 1902.
   3. The *Descent of Man, and Selection in Relation to Sex*. Two vols. John Murray, London, 1871. Ed. used: as above.
   4. The *Variation of Animals and Plants under Domestication*. Two vols. John Murray, London, 1868. Ed. used: 2nd ed., 1875.

5. *The Autobiography of Charles Darwin,* 1809–1882. Edited by his granddaughter, Nora Barlow. Collins, London, 1958. (First published, with omissions, in 1887.)

B. Works by other writers:

1. Sir Gavin de Beer, *Charles Darwin: Evolution by Natural Selection.* Nelson, London and Doubleday, New York, 1963.

2. Michael Banton (editor), *Darwinism and the Study of Society.* Tavistock Publications, London and Quadrangle Books, Chicago, 1961.

3. J. B. S. Haldane, *The Causes of Evolution,* Longmans, Green and Co., London, 1932.

4. Sir Julian Huxley, *The Uniqueness of Man,* Chatto and Windus, London, 1941.

5. Sir Julian Huxley, *Evolution: The Modern Synthesis.* Allen and Unwin, London and John Wiley, New York, 1963.

6. Sir Julian Huxley, *Essays of a Humanist,* Chatto and Windus, London, Harper and Row, New York, 1964.

7. Desmond King-Hele, *Erasmus Darwin,* Macmillan, London, 1963.

8. John A. Moore, *Heredity and Environment.* Oxford University Press (paperback), London and New York, 1963.

9. Wm. Paley, *Natural Theology, or Evidences of the Existence and Attributes of the Deity collected from the Appearances of Nature.* First ed., 1802. Ed. used: James Sawers, Edinburgh, 1818.

10. Leslie Reid, *Sociology of Nature.* Penguin Books, Harmondsworth and Baltimore, 1962.

Index

Index

116

Index

117

Printed in the United States
by Baker & Taylor Publisher Services